D0906597

ENAMEL PAINTING TECHNIQUES

To

My mother whose love, faith
and understanding gave
direction to my career in art

Genisus, enamel steel panel by the author, (1964). 24 in × 33 in (61 × 84 cm).

EDWARD WINTER

ENAMEL PAINTING TECHNIQUES

PRAEGER PUBLISHERS

NEW YORK · WASHINGTON

BOOKS THAT MATTER

Published in the United States of America in 1970
by Praeger Publishers, Inc.
111 Fourth Avenue, New York, N.Y. 10003

Library of Congress Catalog Card Number: 78-107222

Printed in Great Britain

Contents

Foreword

THE ARTIST AND HIS ENVIRONMENT

IN 1932 it was my privilege to introduce Edward Winter, well trained creative artist, to an industrial environment – the Development Laboratories of Ferro Corporation. A graduate and scholarship winner of the Cleveland Institute of Art, and a student of Josef Hoffmann of the Kunstgewerbeschüle, Vienna, Austria, the use of our huge furnaces gave him the idea of extending a traditional jeweller's art to utilitarian objects, paintings and murals for architecture.

Through his talent, industry and skill – his many exhibitions, his two previous books, plus many technical articles featured in trade and art magazines, he has gained a world-wide reputation.

As an American pioneer in the field of glass and metals he exemplifies what can be accomplished by the artist working in an industrial environment.

ROBERT A. WEAVER
Honorary Chairman
Ferro Corporation
Cleveland, Ohio

Acknowledgements

I WOULD like to thank the thousands of persons throughout the world who, through their purchase of my work, have enabled me to devote the time required to write this book. I wish to acknowledge particularly: my wife Thelma whose great talent is evident in the many works of hers reproduced; the Ferro Corporation of Cleveland whose development laboratory at night becomes my working studio; Harold Tishler, for allowing me to use his charming costume jewellery and show readers how it was made; the imaginative enamel of John Puskas; the many architects with whom we have collaborated for the past thirty-five years; Colby College photo by Earl Smith; St Mary's church photos by William Wynne; Crown filtration plant photos by Parade. All other photos in color and black and white are by the author.

Introduction

FOR almost forty years metal enamelling has had a tremendous effect on the art world. This veritable modern renaissance, starting in the USA, has travelled at an incredible pace throughout this country, to England, France, Germany, Italy, Australia, and the Scandinavian countries. Artists and designers discovering it for the first time, or professionals forsaking other media for enamelling, are convinced that no other material can match the range of valuable properties exhibited by enamelled metal.

These qualities include heat resistance, permanence, depth of color, transparency and opacity, as well as texture and design possibilities, not to mention the attributes each artist working in the medium discovers for himself.

The many books and articles written on the subject, including *Enamel Art on Metals*, and *Enamelling for Beginners*, by the author have helped to make enamelling a popular craft. Dozens of new techniques, made possible by processes developed in the laboratory are now made easy for the amateur as well as for the professional.

This comprehensive work reveals for the first time information on the extraordinarily wide technical and expressive range of glass pigments: how to prepare, apply and successfully fuse them on to copper and steel.

A new approach to making enamel copper jewellery is discussed. The cutting and forming of copper animals, birds, leaves, butterflies and their embellishment with opaque or transparent enamels with the addition of silver foil and paillons to add to their luminous depth are all explained. Free form copper shapes, constructions and stabiles are described and illustrated from beginning to end. Steel sectional tiles produced with new drawing and painting techniques have been developed and are described in detail. Glass colors and oxides, metallic lustres, line drawings and washes, and methods of making smelted colors – all fundamental aspects of the new art form – are discussed throughout the book.

The hand tools, brushes, kilns, pyrometric devices and other means used to fabricate and enamel sheet steel into a variety of shapes open an extremely wide field for the craftsman.

The clear step-by-step photographs in both color and black and white were composed and made by the author in his desire to make this involved science easier for the layman to understand. The reader can thus allow his own thoughts, imagination and creative impulses to produce fine art with the highest degree of technical excellence and on a scale never before imagined.

1 What are Enamels?

THE word enamel has come to denote a number of things. It is used to describe any glossy varnish, resinous paint or lacquer of vegetable origin, as well as the coating found on the fingernails of modern females. Many people therefore do not know the difference between vitreous enamels and the numerous glossy substances which share a superficial resemblance.

Vitreous enamel, as the name implies, is a glass, whether transparent, or opalescent, opaque, black, white or colored. When this glass is pulverized, applied in the dry or liquid state to metals, and fused to the base at temperatures of about 1 500°F it is known as vitreous enamel. A range of 300 or so colors and tints is available, comprising all primary and secondary colors, and an infinite variety of intermediate shades and tones.

If a window pane or an old bottle is ground into powder, sprinkled over a piece of copper and fired, the glass will only partly adhere to the metal because this type of glass, being high in silica, is difficult to fuse and cannot be used as an enamel without chemical adjustment. Enamel glasses made for metal are of the borosilicate and lead variety which fuse at lower temperatures.

Durability of Vitreous Enamels

LEONARDO DA VINCI [1452 – 1519] in his notebook writings, *Comparisons of the Arts*, published in English by Edward MacCurdy, devoted considerable space to comparing the permanence of painting to that of sculpture. 'Painting', he writes 'surpassess all human work by reason of the subtle possibilities which it contains.' The one advantage of sculpture is that of offering greater resistance to time. Yet painting offers as much resistance if it is done upon thick copper, covered with a white opaque enamel, then painted on with enamel colors and fired at high temperatures. In its degree of permanence it then surpasses even sculpture. If sculpture in bronze is imperishable, a vitreous enamel painting upon copper may be said to be eternal.

Proof of such permanence is provided by many works of enamelling art from the fifth century BC which are still perfectly preserved in European and American museums.

Glass – An Age Old Material

MILLIONS of years before the beginning of recorded history, quartz, one of the Earth's basic materials, was fused into a form of glass by volcanic heat. It is perhaps paradoxical that though glass in its modern forms is one of the most amazing materials of twentieth century technology, man has been making useful and beautiful objects from it for 3 500 years. By the application of human skill and through the abundance of glass-making materials over much of the Earth's surface, the usefulness of glass has been greatly extended, so that today it plays a vital and ubiquitous rôle in our lives.

Technical debate often takes place as to what glass really is, and there are many theories about its structure. In several ancient civilizations, glass making was one of the refined technological arts. How did the ancient glass makers produce their glass, and under what circumstances did men first make it? Glass was used in the form of beads and glass vessels in Egypt early in the eighteenth dynasty [1580 – 1858 BC]. Since that time many hundreds of glasses each with their characteristic properties and chemical composition have been formulated.

Glass has always had built-in attributes of strength, clarity, brilliance, depth, transparency or translucence, though like all man-made materials it has its faults: it is very brittle, for instance.

It is only within the last hundred years that man has begun to sense the almost limitless possibilities of glass as a material. Technologists of this country have developed glasses strong enough to drive nails, withstand bullets, resist sudden and violent temperature changes and stand up to the corrosive action of strong acids. Other glasses can be woven and made into fabrics, applied and fused on to cast iron, sheet metal, and aluminium. Special glasses designed for the space age are helping to send astronauts into space to explore the planets. Today, by using the information in this book, the contemporary artist can learn to paint with glass in a finely ground state or fuse it in lump form to metals.

2 Glass Forming Materials

SILICA forms the basis of all glass. It occurs in nature as quartz (flint, sandstone, and certain other forms). Tertiary sands lying on secondary beds are the common raw materials for glass making. If they originate from quartzites they are exceptionally pure. Sands from weathered granite usually contain undecomposed feldspar and kaolin. Any mica present can injure the quality of the sand but iron impurities can be most injurious owing to their coloring or darkening effect on the glass. Through processing and refining, any foreign matter can be eliminated, and pure quartz sands are readily available. It is normal to grind the sand very fine for use in borosilicate glasses.

One of the difficulties encountered in making glass is the high melting temperature of silica : about 3 110°F. This problem is solved by adding fluxes to the silica to lower its melting temperature. Typical fluxes include lead, sodium, potassium, calcium and manganese oxides, borax, fluorspar, and many others. Lead crystal, very beautiful and costly glass, is still made for expensive tableware and decorative items, but lead is not now so widely used in ordinary glass making because of its cost. Glasses made for fusing on to metals such as gold, silver and copper are fluxed with relatively cheap materials such as potassium and sodium silicate, borax, and sometimes specially prepared frits based on lead bisilicate.

BORIC OXIDE

Boric oxide, B_2O_3, is widely used in glass and frit manufacture. Its use permits high silica concentrations in the glass at relatively low fusing temperatures.

Borax, sodium borate, can be successfully used in medium to high temperature glasses but in the jeweller's or painter's type of glass the sodium content of plain borax would be injurious, so boric acid is used. Boric acid volatilizes at temperatures of 1 000°F and above, so care must be exercised in firing and fusing boric compositions.

SODIUM OXIDE

This material is introduced into a glass batch as calcined soda (soda ash) and is responsible for some of the brilliance in the glass; however, too much will increase the coefficient of expansion and lower the resistance to thermal shock.

POTASSIUM OXIDE

Potash, K_2CO_3, is used extensively in lead or borosilicate glasses and a considerable amount is used in the ruby-type glass. Potassium has a considerable effect upon increasing the brilliance of the glass and it has a more favourable effect on the physical properties of the glass than does sodium oxide. It does not increase the expansion coefficient as much as soda, but the elasticity of the glass is increased rather more, giving greater resistance to thermal shock.

MAGNESIUM OXIDE

This is used in dry process glasses specifically for cast iron and in small amounts as a mill addition to alter certain other properties of the enamel.

BARIUM OXIDE

This is occasionally used to increase elasticity in lead glasses. Its one advantage over lead is that it is not susceptible to reduction by furnace gases or by iron. It is highly toxic and is not used widely by craft enamellers for this reason.

LEAD OXIDE

Lead oxide is the ideal material for low temperature glasses since it possesses the highest refractive index and confers greatest brilliance on all glasses. Its expansion coefficient is low and its elastic properties favourable (helping the enamelled metal to resist thermal and mechanical shock). Barium sulphate is sometimes used as a cheap substitute.

ZINC OXIDE

This material is also used in white glasses designated for dry process preheated cast iron. Its elastic properties are not very good but its resistance to chemical attack is.

ALUMINIUM OXIDE

Alumina, Al_2O_3, is present in most glasses. It confers hardness and helps to reduce devitrification or crystallization tendencies. It is contributed either by commerical alumina or more commonly by feldspar, kaolin and other glass-making materials.

TIN OXIDE

For many years tin oxide was the principal opacifying agent in glass. Today zircon, antimony and titanium oxides are used, since tin oxide is very costly. For certain unique effects tin is essential. The way in which these agents opacify enamel glasses is quite complex. Simply, the fine particles of opacifier scatter the light and give an opaque effect to the medium in which they are contained.

3 Classification and Definition of Glass Enamels

TRANSPARENT ENAMELS

A GLASS that allows light to pass through it is called transparent. Some types are completely clear. Transparent colorless glasses are classified as light, medium or dark depending on the amount of pigment contained in the glass. The underlying design, line or pattern will show through to various degrees, depending on the color depth of the glass fused on top. Polished metal surfaces will show through to the same degree.

OPALESCENT ENAMELS

These are similar to transparent glasses but there is some cloudiness or turbidity, giving the impression of a semi-opaque milky veil. This subtle tonal glass produces a surface of delicate charm. Too high a firing temperature or too many firings will often kill the semi-opaque transparent effect. Timing and firing must be thoughtfully controlled.

OPAQUE ENAMELS

A solid and compact glass through which light cannot pass, but which can be reflected from the surface of the glass. The opacifying agents are built into the glass during melting to produce millions of minute particles which prevent light transmission.

TRANSLUCID ENAMELS

The hard or soft clear glass fluxes are classified in this section. Extremely light pinks and yellows sometimes fall into this group.

IRIDESCENT ENAMELS

Metallic lustres including gold, silver, platinum, copper and mother-of-pearl produce a light iridescence when painted or sprayed on to a surface of white fired enamel. Other colored opaque and transparent base glasses can serve as a suitable base for lustres.

PHOSPHORESCENCE (FLUORESCENCE)

Glasses made with the use of pigments such as zinc sulphide, calcium sulphide, willemite and strontium sulphide are called luminous pigments. These give off glowing effects and assorted color tones when subjected to ultra-violet radiation. Some luminous pigments contain radium and will glow without ultra-violet light, comparable to watch and clock faces that glow in ordinary light.

SATIN-MATT

This is a glass that has no gloss or shine and allows no reflection. It is produced by grinding a high percentage of refractory material into the glass or by controlled crystallization procedures. Loading a glass batch with silica or alumina or combinations of titania and zircon will produce this effect. Calcium and zinc matts are also available. These minute-tooth surfaces are ideal for drawing or brush painting.

White Enamel as a Base for Painting

ALMOST all the examples of art enamelling illustrated in this book are paintings produced on a white enamel surface. A white base gives the artist the advantage of maintaining accurate values with all transparent colors fired over it, and which still show no fading or burning out when repeatedly fired.

The four oxides most generally used in making an opaque white glass are lead, titanium, zirconium and antimony. The white opacity due to crystalline particles of these oxides distributed in the glass, either during the glass melting process or by milling them into the glass, depends on the quantity, particle size and refractive index of the crystalline particles. The thickness of the coating also controls the degree of whiteness. The tone of whiteness, cool or warm, depends upon the character of the crystalline material and the color of the glass which surrounds it.

Titania is considered the best opacifier for whiteness. It also confers resistance to chemicals and acids that may corrode the glass. Titania, zirconia and antimony oxide are usually melted in a clear glass but in most cases tin oxide is ground into the glass in a ball mill to produce opacification and whiteness.

All of these opacifiers will produce soft, medium-hard or hard glasses depending on the composition of the glass itself. White opaque enamels for steel have a wide firing range, from 1 000° to more than 1 600°F. The type with which we are concerned fuses at 1 400°F. The transparent and opaque enamel colors which have been compounded to melt usually at lower temperatures are fused upon this prepared surface.

Enamels with Matt (No Gloss) Finishes

MATT enamels with their dull and extremely fine-tooth surface will prove to be one of the greatest assets to the enamellist. It is very difficult to paint on smooth glossy surfaces since the enamel tends to flow too freely on such surfaces. In the past some artists tried dulling the surface with hydrofluoric acid but this was not completely effective, and the use of this acid is hazardous. Sand blasting was sometimes used, but this left the surface too rough. A matt enamel can be produced by including a relatively high percentage of refractory materials such as silica, zinc, zirconia or alumina. In amounts of 1 to 3% any of these will reduce gloss. Combining two dissimilar enamels, e.g. a soft highly fluxed type with an extremely hard refractory, will also produce a matt surface. The firing range of this type of enamel is from 1 250° to 1 560°F. The acid resistance of the surfaces produced may be as high as high-gloss enamels.

The color range of matt enamels is somewhat more limited than that of the gloss enamels which is almost limitless. Brilliant red, orange or yellow colors are most difficult; blue, olive, green, brown, grey, pink and a variety of earth tone colors are easily made and come in various degrees of matt, including semimatt, semigloss and dull matt.

Many examples of painting and drawing in this book were executed on matt surfaces.

4 Painting

MOST successful artists will readily admit that drawing is a prerequisite to painting. One artist friend admitted recently, 'I have discovered how very much integrated my drawings are to my paintings; to paint without drawing experience is like making love by correspondence'. To this I might add, a knowledge of design is as equally important to the artist as a knowledge of drawing and painting. The fortunate individual possessing an abundance of these is on the way to becoming a master.

The tactile appeal of fired semitransparent, opalescent enamels, swirling into metallic streams and puddles, awakens an almost uncontrollable urge to fondle the surface and possibly ask: what material means has released this spiritual depth that seems too subtle for the eye to perceive?

In further extolling the qualities of fired enamels I am reminded of a quotation attributed to Frank Duveneck, one of our late nineteenth-century American master painters schooled in both Europe and the United States: 'Painting mediums and the tactile appeal constitute a lust that materially heightens aesthetic response. Rembrandt had it in the fatty impasto of his stand oil mixtures; Degas and Renoir would have lost much of their charm had they not appreciated its value. Cesanne got it by patiently building layer upon layer of paint until it resembled fine old *cloisonné* enamel whose whites are never the stark white of tube paints, but the milky, vibrant, iridescent concoctions of an alchemist – the fire giving it color indescribable'.

With these new drawing and painting techniques the student should no longer feel frustrated in his efforts to control and master enamelling. This new freedom is like loosening the chains of tradition so that almost anyone can experience the thrill of accomplishment with this all-purpose art medium.

The four or five traditionalist enamelling techniques were founded and developed by artisans who were severely limited by their materials; the metals were impure and crude, and there were no metal standards upon which one could rely in striving for perfection. The enamels were smelted from secret formulae in a hit-or-miss manner; the firing was precarious, for the heat was either too intense or too gentle. Under extremely hot melting or firing the enamels would run uncontrollably; there were no pyrometers to control the temperature. Owing to such adverse conditions, *cloisonné* and *champlevé* techniques continued to be fostered and remained popular. The small flat wires of *cloisonné* enamelling were necessary to keep the colored enamels confined to a set area, rather than run and disfigure the work. In the *champlevé* method a heavy metal was used so that cells could be gouged out which would serve to hold the enamel. The extremely heavy gauge of the metal made objects more durable for constant use, and served to keep the enamel from breaking off the metal.

There is no turning back to the ways and glories of the past. The speed with which science and art have joined forces in our time is startling; in fact, even our comprehension of the aesthetics of painting and sculpture is insufficient today to keep pace with the tremendous impact of the new materials discovered and produced by science and industry. Today one of the greatest challenges facing us all is the need to acquire a better understanding of materials from the beginning of their formation, their chemical composition and molecular structure. The new uses to which we put an ancient material such as glass enamel, demands imaginative thinking from both artist and scientist.

Glass Fluxes for Enamel Painting

GLASSES made for painting are usually compounded to fuse at moderate temperatures. The very soft enamels fuse from 900° to 1 100°F, and the medium type at 1 100° to 1 350°F. In practice the enamel paints must fuse at a lower temperature than the surface upon which they will be applied. The base coat thus remains stable and is hardly affected even by repeated firing of the colors applied to it.

COMPOSITION AND SMELTING

Quartz sand (silica), red lead, borax or boric acid are the usual ingredients of a clear glass flux. The properly constituted mixture is ground to a fine powder in a ball mill. It is then placed in a red-hot fireclay crucible which is placed in the furnace for melting. When the glass has been thoroughly melted the crucible is picked up with large tongs and the contents poured into a vat of clean cold water. Sometimes the molten material is poured over large slabs of steel for cooling. Glasses containing high percentages of boric acid are usually poured on to cold steel, since water used in cooling might leach out some of the soluble borates. Having been granulated by sudden quenching or cooling, the frit must next be reduced to a fine powder by grinding in a ball mill. Sometimes the larger lumps of glass that form when it is poured on to steel slabs are crushed in a steel crusher before being milled.

To change this clear glass into the desired colored form the glass must again be melted with the proper percentage of pigment or coloring oxides. In producing colors of light and dark tones a graduated test sampling is carried out.

Test No. 1 : 90 parts coloring oxide
10 parts glass flux
Test No. 2 : 80 parts coloring oxide
20 parts glass flux
Test No. 3 : 70 parts coloring oxide
30 parts glass flux

This procedure can be continued until No. 9 mix is prepared, consisting of 10 parts of color and 90 parts of glass flux.

These color samples can be painted over a white enamel surface and fused, giving an accurate indication of the tonal differences. This process can be repeated with several other pigments to produce the colors desired. Most enamel suppliers stock large quantities of clear glass flux, and simplify the manufacture by melting up colored frits as they are ordered. A range of clear fluxes is usually available for different purposes and metals.

Colors capable of withstanding higher firing temperatures, such as cobalt blue, iron reds and pure chrome greens, are used with highly siliceous fluxes. Colors ranging from pink to dark red purples are usually obtained from the gold ruby or copper ruby glasses; they will not withstand high temperatures without losing their peculiar quality. In these instances the glass batch must be mixed with fusible fluxes such as boric acid, replacing much of the silica. This is known as a lead borate glass and it will produce delicate shades that will maintain their true color after firing.

Painting Media

RESINS

There are two types of coniferous resins: soft damar and hard copal. They are both fossil materials obtained from the petrified sap of certain kinds of needle trees extinct since 7 000 BC that found its way, like other ancient materials, into the deep interior of the Earth. In early times this amber material was used in the jewellery trade owing to its lustrous color and easy working qualities. The chief source of supply is the Baltic coast, but it is also found in Canada, England and other European countries. The hard variety of resins were at one time called 'glassa' since they were almost vitreous in appearance. Soft resins like damar are easily dissolved in turpentine or mineral spirits and the smooth working material dries into a thin film. The film burns away when placed in the heat of the kiln.

THINNERS

The best turpentine is labelled 'pure gum turpentine' or 'gum spirits of turpentine'. It is sold in paint stores, hardware stores and in some art supplies stores. Gum turpentine is distilled from thick viscous oleoresin obtained from pine trees grown in parts of the USA and Canada. The balsam tree is another source of this very thick sticky material. Viscosity modifiers are numerous, but ethyl cellulose is most popular as a thickening agent. The average frit-oil ratio used is about 80 parts of glass (enamel): 20 parts of oil. A heavier working material is obtained from an 85 : 15 ratio. Thinner washes can be produced by using a more viscous oil in the ratio of 70 : 30.

APPLICATION BY SPRAYING

The most widely used spray medium is alcohol and water, generally made up as follows: 50 parts by weight of enamel, 15 parts of water and 5 parts of denatured alcohol. Some of the water can be replaced by more alcohol to increase the drying rate; conversely more water can be added to the alcohol to produce a slower drying rate. A mixture of turpentine and fat oil is sometimes used for spraying when a heavy application is required. A typical mixture would be 6 parts of enamel, 1 part of squeegee oil (used in silkscreening) and 1 part of turpentine. A squeegee paste combined with lacquer thinner will also produce a satisfactory spraying medium. White shellac, when diluted with four or more parts of denatured alcohol, makes a good spraying or brushing medium.

SPRAYING EQUIPMENT

Many types of air brush and spray gun are available. The small air brushes require the enamel to be extremely thin. All the thin colored lustres, and gold, silver and platinum preparations are easily applied with this type of brush activated by compressed air. A pressure of 30 to 40 psi is adequate and can be used to apply all types of enamel media, both of the oil and of the water variety.

Brushes, Pencils and Tools

BRUSHES for enamel painting are as varied as those used for oil or water color. Camel hair, sable or bristle can be used successfully for each specific requirement. The standard pointed sable brushes in small, medium and large sizes are ideal for detail, filling in areas with color or inlaying washes. The liquid lustre colors handle well for all wash effects.

When defining small details where circumspection and precision count most, such as in portraiture, the small pointed brush is indispensable, since it allows maximum control and at the same time prevents the artist from becoming too impulsive. The bigger brushes, including the round and flat edge types, can be used for painting proper and for covering more area with the same craftsmanlike delineations as with other art media.

The extremely long-tipped sable brushes are instruments of incomparable properties, and these can become seismographs of sensitivity. The long, blunt edge script liner takes up a quantity of paint and can produce strong as well as split hair-lines in great quantities without the need for returning to the jar for replenishment. If the medium is to work well with this brush it must be fairly thin and possess good stringiness and tackiness.

While the sable and camel hair brush can be considered the supreme brush, with no limitations of any kind imposed on the hand that moves them, the pig bristle brushes (those with flat ferrules) and in graduated sizes, can perform a fine job in applying paint.

Since pig bristles are chisel shaped and coarse compared with the sable type, the application of the medium to the surface will reflect this coarseness. These brushes are excellent for drawing in compositional lines.

The stub end stencil brush in various sizes is useful when applying color to a confined area, protected with paper template. The wide 3-in and 4-in (76 and 102 mm) brush is best for quickly covering large spaces.

The successful handling of brushes, making them perform his bidding, and getting the desired results in the application of the medium, rests with the practice and ability of the artist, his personal devices, intuitions and compulsions.

One discovers that by allowing the wrist to swing freely things can be made to happen rapidly and auspiciously on the painting surface; brush strokes become the semaphores of a heightened artistic sensibility. The nervous, yet controlled movement of one's hand expresses itself through the medium of strokes that intuitively follow its dictates, for a brush stroke can be as eloquent a proof of authorship as a painter's signature, and the touchstone of his virtuosity.

PENCILS

By using white, black, grey and assorted color matt surfaces, the artist increases his range of enamel drawing surfaces. Prior to the appearance of matt finish, the high gloss surface of most enamels was too smooth and slick to receive a line drawing. Drawing and sketching potential is an aid to the painter in preliminary composition and design layout. The flat or graphite pencil, sharpened to a chisel edge, will make both a fine and thin line on the white, grey or colored matt enamel surface. A series of four pastel colored pencils is now available : yellow, pink, green and grey, to add further possibilities to this new medium. The white drawing pencil introduced several years ago, being highly effective on dark or black surfaces, is very popular. The paper stomp widely used in chemical drawing, or the finger, can be used for tonal gradations. These pencils are obviously made of a material that will not burn away when the work is fired.

TOOLS

All sorts of metal tools and wooden sticks can be used to produce unusual effects. The razor blade, the dental tool, the pocket comb and serrated knives can be used for graffito techniques (scratching through a dry coating of enamel to a base of another color). Pointed hardwood sticks, brush ends, or toothpicks slide over the enamel surface easily, and produce a good line. Extremely fine lines can be obtained with the use of a dental tool.

Intermixing Glass Pigments

THE following practical rules based on the author's wide experience of enamel blending are offered as a general guide.

1. Do not intermix colors having different firing ranges, such as those fired at 1 050°F with those fusing at 1 300° to 1 350°F.

2. Transparent colors can be lightened in value by the addition of a small amount of white; however, this also makes the color opaque (as in the case of adding white tempera to transparent water colors).

3. Transparent colors can be darkened by the addition of black but this may opacify the color, and in some instances radically change it: black added to lemon yellow will fire with a greenish cast.

4. Transparent colors can be respectively lightened and darkened by the addition or subtraction of a clear flux or a darker color of the same glass.

5. The cool colors (blue, green, violet and crimson) intermix well with each other; and the warm colors (red, red-orange, orange and yellow) have a like affinity, but the two types do not work well if intermixed.

6. Transparent colors fired above their designated firing temperatures will tend to fade or lose intensity. To maintain a set color value two coats may be necessary with a firing between the two.

7. To maintain subtle values of color great care must be taken with firing and proper timing and accurate pyrometer temperature control exercised.

8. Enamel suppliers have a wide color range containing several hundred color values. Suppliers' color charts are available to save time on workshop blending.

9. Dry glass powders can be applied to a painting by means of a brass hand sieve (200 to 250 mesh). This technique is fully explained in the author's book, *Enamel Art on Metals*.

10. Opaque enamels ground in water and clay can be intermixed successfully. Using a medium grey or brown, a wide range of light values can be produced by adding white, or darker effects can be obtained by the addition of black. Warm tones can be acquired by the addition of yellow or orange, and cool colors produced by the addition of light or dark blue.

11. The pink, rose and red range of colors is made with gold salts; these colors intermix well in powder or paint form.

12. It is recommended that the artist makes color test samples (transparent as well as opaque) on a white enamel sheet to discover the possible color range, and the reaction of one color with another after firing.

13. This book deals with the execution of fine art and goes beyond the limited craft of china painting. The imaginative, experimenting artist is free to break all rules restricting the glass technician. Coloring oxides and glass pigments subjected to intense heat have unlimited expressive potential.

5 Colored Lustres as Painting Media

THE painter will find great satisfaction in painting with lustres. These inorganic (mineral) pigments are highly iridescent and can be applied on a glass enamel surface in much the same way as water color washes. Although these substances have the consistency of heavy inks they can be applied with a camel hair brush, by airbrush or by screening.

COLOR POSSIBILITIES

Colored lustres produce a metallic sheen in addition to a thin coating of color on the surface. They can be classified as colorless and colored. The former is a preparation in which the white-splashed surface upon which it is applied takes on a mother-of-pearl effect. Colored lustres impart a metallic lustre in addition to the color itself, making it possible to acquire effects similar to oxidized bronze, copper tones, and the iridescent coloring found in the plumage of birds, especially peacocks' feathers. An assortment of interesting effects can be produced by combining both types of lustres in varying proportions.

Lustres are obtained by reacting metallic oxide with ordinary pine-oil resin, dissolved in a suitable solvent such as lavender oil. The amount of oil determines the viscosity of the solution.

The colorless lustres are: alumina, bismuth oxide, lead oxide, and zinc oxide lustres. The colored lustres are: copper, cobalt, nickel, uranium, cadmium and iron.

PREPARATION OF WET LUSTRES

The main ingredient used is a solution of sodium resinate or soda resin soap produced from an aqueous solution of sodium carbonate which is heated to boiling point and treated with small amounts of powdered pine resin. The resin consists of acids which are powerful enough to displace carbonic acid from the carbonate part of the metallic compounds normally used to make resinates. When the resin stops reacting it is a sign that the remaining solution contains nothing but dissolved sodium resinate. This liquid, after dilution with water and boiling, is allowed to settle at room temperature. The clear solution, which is pale yellow in color, is ready for use as a metallic resinate liquor. Many of these resinates have a tendency to lose their solubility in essential oils upon prolonged exposure to light, and should be stored in opaque glass bottles.

LUSTRE MADE WITH GOLD

The most luminous lustres can be made with liquid bright gold, commonly used for decorating ceramics, in varying proportions with bismuth lustre. Five parts of liquid bright gold and one part of bismuth lustre will fire out a copper color; and if the proportion is raised to two or three parts of bismuth lustre to one part of gold, the effect will be a bluish-violet shade with a golden sheen. If the gold amount is increased the blue changes from purple to rose. Numerous tonal effects can be produced by adding lustres containing iron, chromium, uranium, nickel, etc. to the gold-bismuth lustre mixture. Variegated tonal effects can be made by a series of downward brush strokes, allowing the different colored lustres to run partially together.

Gold chloride, stannous and stannic chloride when combined in certain critical proportions will give the well-known color purple of Cassius. The purple shades will again depend on the ratio of each compound used. A gold-tin combination of 1 : 10 will make a maroon color, 1 : 5 a rose, and 1 : 4 a light purple shade.

PLATINUM LUSTRE

This silver-toned lustre is often used in place of silver because it does not tarnish, as silver does, under the action of sulphur from the atmosphere. It can be prepared by diluting liquid bright platinum (used in pottery decorating) with lavender oil and nitrobenzene or by dissolving a saturated solution of platinum chloride in lavender oil. Extreme care must be taken with nitrobenzene which is highly toxic. This solution has an excellent covering power if not overdiluted. If applied and fired on to a white titanium glass surface it will fire in a solid coat; if, however, it is painted on to zircon or lead compositions it will tend to crackle (see Glossary). These compositions tend to mature at lower temperatures. It is advisable for the artist to experiment with this lustre on all types of glass and keep accurate temperature and timing records so as to be able to repeat his results.

SILVER LUSTRE

Silver lustre is made by mixing silver nitrate with lavender oil and firing in a chemically reducing atmosphere (that is, when the kiln is starved of oxygen). As with all other metallic coatings, resinous oils must be smoked off and fired carefully. If after silver lustre has been fused on to glass enamels the surface becomes grey or dull, it can be polished with buffer and metal polishing or scouring powder. As with all other lustres, the production of a solid surface or crackle effects will depend on the type of glass over which it is fused.

By adding 2% silver carbonate to a transparent colorless lead glass and firing at a low temperature it is possible to make one's own silver lustre. A yellowish silver lustre is obtained by mixing silver chloride with three times its weight of clay and ochre and sufficient water to form a paste. Other lustre colors can be made simply by mixing silver chloride or nitrate with fat oil, lavender oil, or with nitrobenzene or honey. This combination will give off a greenish tint with the faint suggestion of gold. The very wide color range of lustres and the ease of making and applying them (using a pointed brush) offers the artist an exciting liquid medium.

SOME LUSTRES AVAILABLE FROM SUPPLIERS

Mother-of-pearl	Bright orange
Iridescent light blue	Rose pink
Turquoise blue	Blue
Dark blue	Blue violet
Purple violet	Amethyst
Sapphire	Pink
Burgundy (old rose)	Chamois

6 Airbrushing

AIRBRUSHING LUSTRES

SUCCESSFUL airbrush decorating with lustres requires a good deal of experience with the use of the airbrush as well as familiarity with the use of lustres. The beginner would do well to experiment with water color and paper in order to get the feel of the brush. The scope and problems inherent in the use of lustres should be thoroughly studied.

A double-action airbrush has been found to give the best results. Experience in dismantling and assembling the airbrush is essential to the artist because it must be kept in perfect condition to do a satisfactory job. Gold enamels and lustres have a tendency to creep into the mechanism of the airbrush (especially into the part to which the air hose is attached) thereby clogging the action. The airbrush must be thoroughly cleaned after use; once the lustre has dried the brush is very difficult to clean. Lacquer thinner is suitable as a cleaner (never use alcohol for this), and following the cleaning be sure to oil lightly the parts of the brush as instructed by the manufacturers.

Lustres work somewhat differently from other spray media in that they dry very quickly upon application; this is an advantage, since otherwise dust settling on the wet colors and absorbing moisture from the air would result in white spots that become visible only after firing. Spraying should not be attempted on a very damp day, and the piece to be airbrushed should be warm (but not hot) in order to hasten the drying. The worker must also guard against coughing, sneezing, and putting fingermarks on his pieces if defective ware is to be avoided.

Lustres are quite flammable during spraying and therefore the job should be done always in a well-ventilated room. A clean spray booth is useful if this can be made available. Never airbrush lustres near a hot kiln or an open flame. Most professional artists use a thin plastic stencil film, available from most artists' supply stores in sheets or rolls.

Whatever type of stencil material is used the piece should be cut large enough (with a sharp knife or razor blade) to protect the areas of the design that are not to be colored. If some lustre creeps under the sharp edge of the stencil it is possible to use a light coating or rubber cement on the underside of the stencil so that it will hug the surface. Both the positive and negative sections of the stencil can be used. This yields sharp contrast and confers a third dimensional effect on a design. Mottled spotting, light textures and other disfiguring marks in the lustre can be achieved by breaking the above rules and applying drops of water or alcohol to the unfired surface. Quite unusual circular shell spots can be obtained by dropping ethyl acetate on to the surface with an eye dropper. Artists can experiment with many materials to discover other unusual effects in the fired lustre. The temperatures at which lustres are fired are 1 300° to 1 350°F for about three minutes. Higher temperatures produce burnt-out surfaces or other unattractive effects.

7 Drawing

DRAWING has always been an important part of any early school curriculum. Children at all levels can easily experiment with drawing and painting on tiles or other suitably prepared enamel work surfaces. Owing to the popularity of enamelling many schools already have small kilns, possibly for art potting, if not for enamelling. Students' sketches, drawings and free brush paintings can therefore be rendered permanent by a three-minute fusing.

This type of enamelling is flexible enough to be used and intermixed with many different educational techniques, and this book should satisfy the demands of teachers and students since the working rules are explicitly set down. Design, technique and vocabulary are most simple, and the limitation and potentials of the craft can be explored as they arise. Most students are anxious to learn practical procedures, and some will undertake the necessary steps to achieve their goals. White, black and colored drawing pencils have now been produced for use in enamelling, and there is every chance that a small project successfully executed will serve to direct a student into further study and to a successful fine art career.

GRAPHITE PENCIL

The graphite pencil provides the enamellist with another fine technique – the ability to draw with rich black line on a white or light matt enamel surface.

Graphite is crystallized carbon. It occurs free in nature and is also made artificially by heating finely powdered Pennsylvania anthracite coal in an electric furnace. This material is exceedingly inert and will withstand very high temperatures, and still remain graphite. Graphite is mixed with clay (free from grit) and ground in water between millstones and then squeezed through a die in the form of a thin rod. This is then dried and fired at a high temperature to toughen it. When cool, the rods are enclosed in wood to make pencils. The more graphite in proportion to clay, the softer and blacker the pencil. With practice the student can soon learn to produce tonal drawings that are light, medium or dark by the manner in which the pencil is used. Cross-hatching is possible, and solid black areas of a design can be built up by filling the areas with heavy strokes of the soft pencil.

Luminous transparent glass enamel in lump (frit) form has a jewel-like quality. It can be purchased in this manner from the supplier. It also comes in dry powders for sifting and painting.

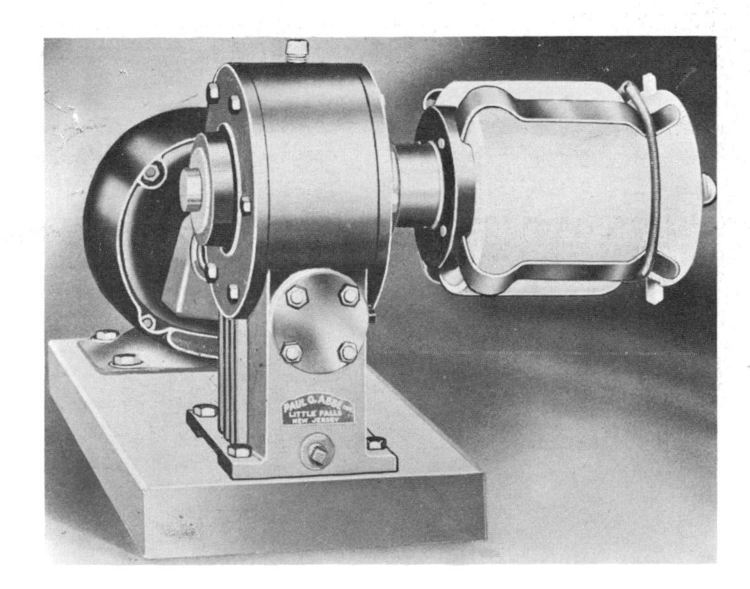

Single Bascilla ball mill ideal for school or studio, is equipped with a $\frac{1}{6}$ hp/115 volt motor.

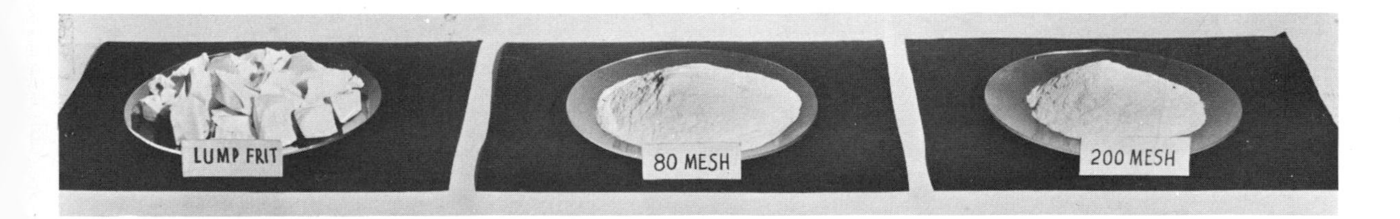

Large lumps of glass enamel are produced by pouring molten glass on to huge steel slabs to cool. It is broken up with a hammer. The 80 mesh enamel is made from grinding lump frit in a ball mill. The fineness depends upon the length of time it is allowed to grind. Fine enough to pass through an 80 mesh sieve. The 200 mesh and 250 mesh enamels are extremely finely ground for suitable painting media.

PLATE 2

A steel pestle and mortar can be used to pulverize small amounts of enamel, the contents of the mortar poured into the sieve and shaken on to clean paper.

Molten glass enamel pouring from a 2 400°F smelter and into a tank of water. Particles are called frit.

Feldspar rock from which silica and other glass making materials are derived.

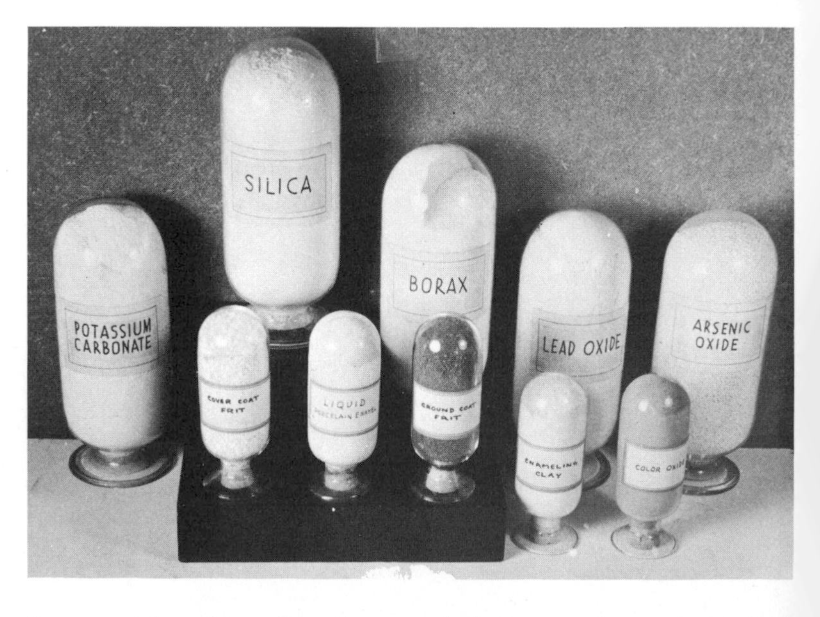

Raw materials used in making glass enamel. Potassium carbonate, borax, lead oxide and arsenic oxide. The small bottles show the glass frit, liquid enamel, ground coat frit, enamelling clay and color oxide.

Colors remain constant after firing over white coated steel plaques.

Colors ready for painting can be placed on palette or if larger quantities are desired in a cake tin.

Enamels for painting come in a wide assortment of colors. The dry powdered form is most popular.

Applying liquid gold to the design.

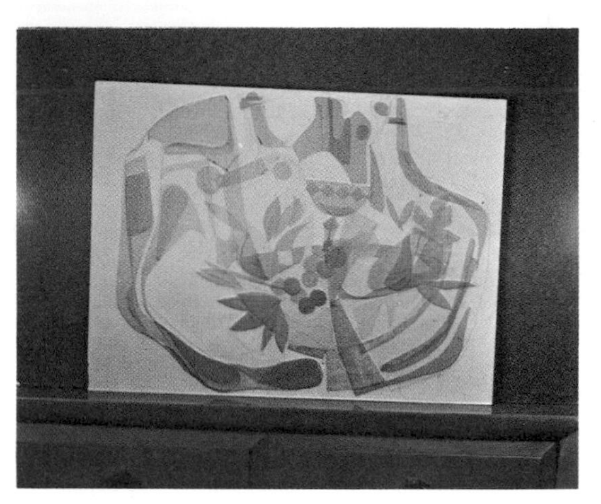

Thin sheeting comes in a variety of colors.

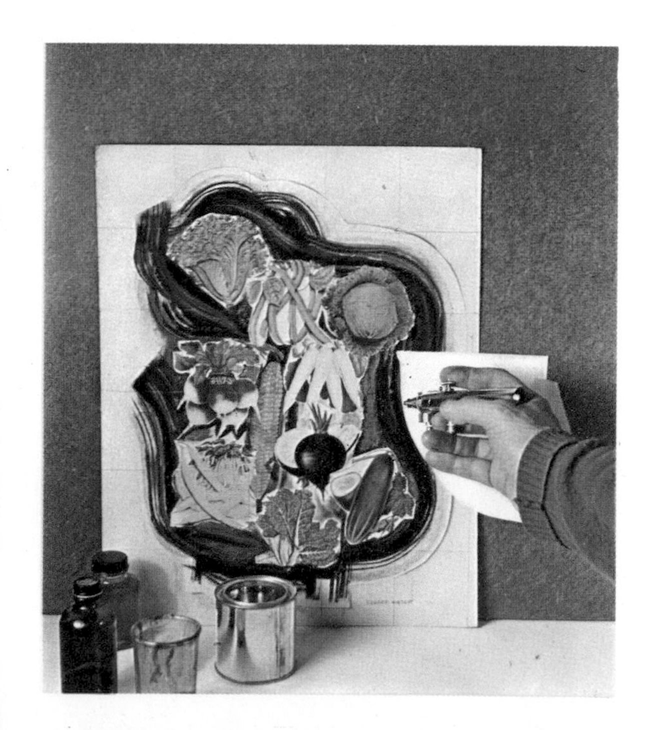

Finely ground enamel (250 mesh or finer) can be applied with air brush and compressor.

Oil-base screening enamels are manufactured for all types of silk screen printing. Silk is used for small runs and fine wire screen for mass production. Silk screen enamelling is an ideal school project.

PLATE 6

Screening type enamels in an oil base are processed for use as printing media.

Oxides are weighed out in proper proportion for screening paste.

A pulverizing machine is used to grind metallic oxides that have been purified through intense heat.

Detail is painted with a pointed brush.

Stencil brush used for pouncing color through template.

Sunflower serves as motif for an enamel plate design.

PLATE 8

Plates and box inserts in colorful designs by Thelma Frazier Winter.

Leather boxes with enamels are an ideal combination.

Sensitive and delicate renderings can be accomplished by the use of the air brush.

Fruit composition.

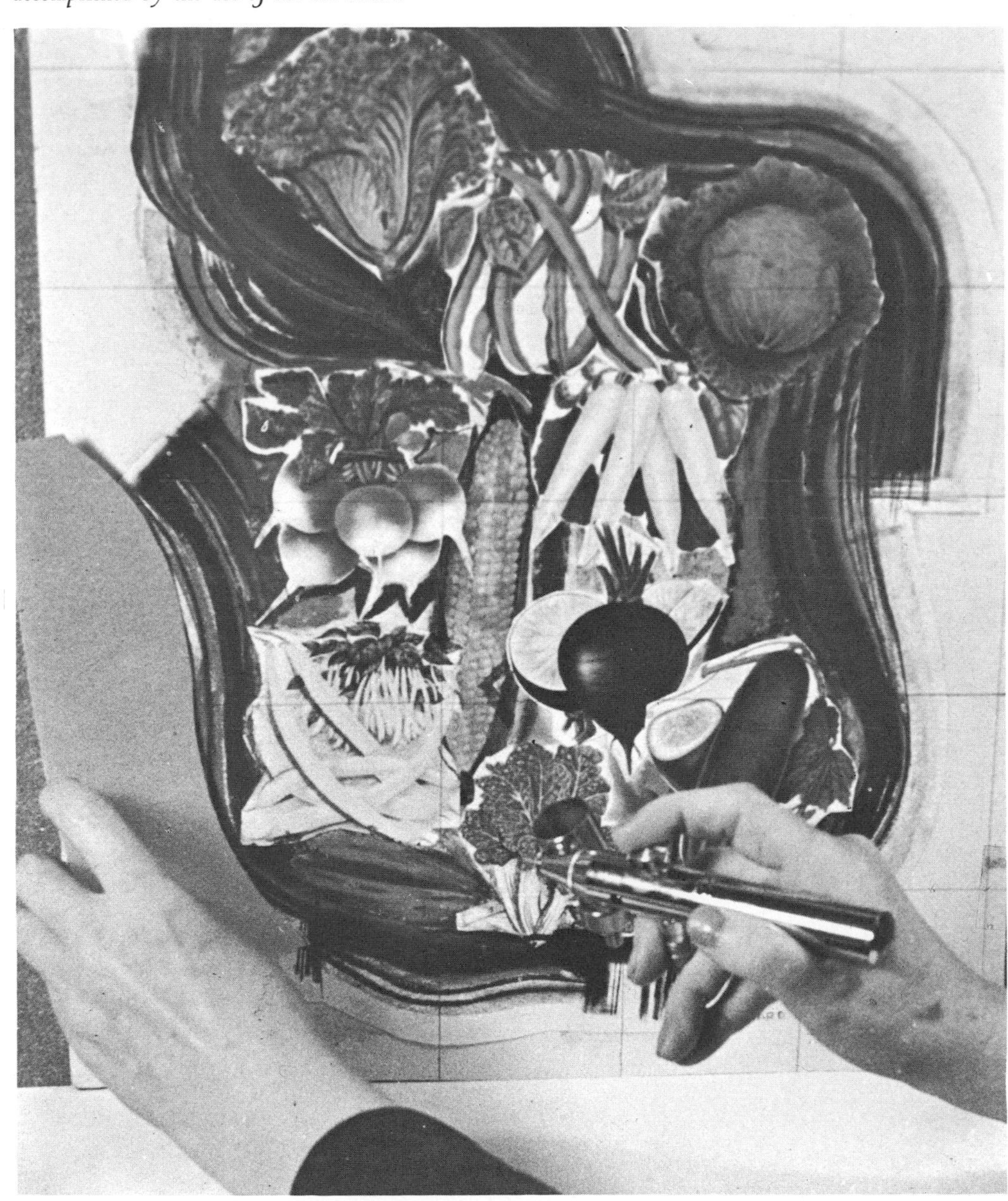

air brush and pressor can be used to y enamel. A paper ight-edge template is to protect areas.

PLATE 10

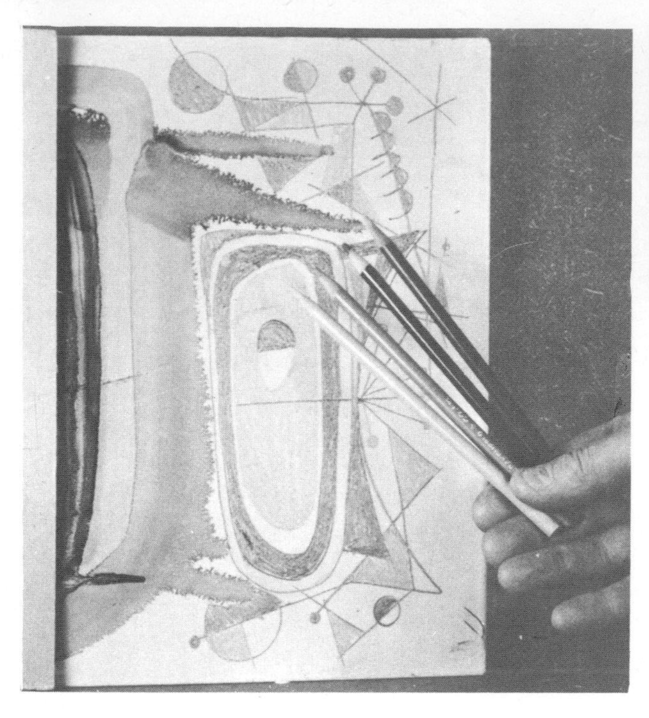

Underglaze ceramic pencils are used to produce yellow, red, green and grey drawings on a white matte enamel surface.

The black line drawing overlaps several white square tiles.

Shading a line drawing.

The finger is used to tone the drawing.

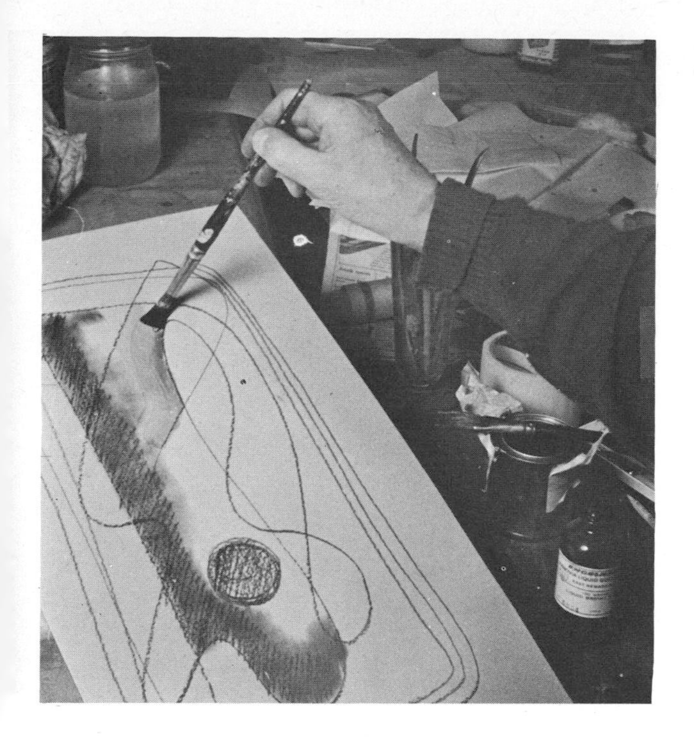

A purple lustre is washed over the line drawing.

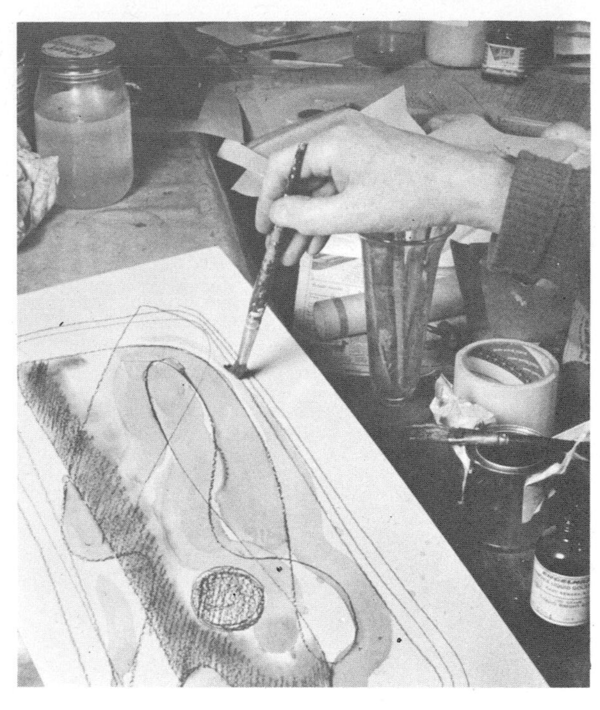

Quick action with the brush will produce effective washes.

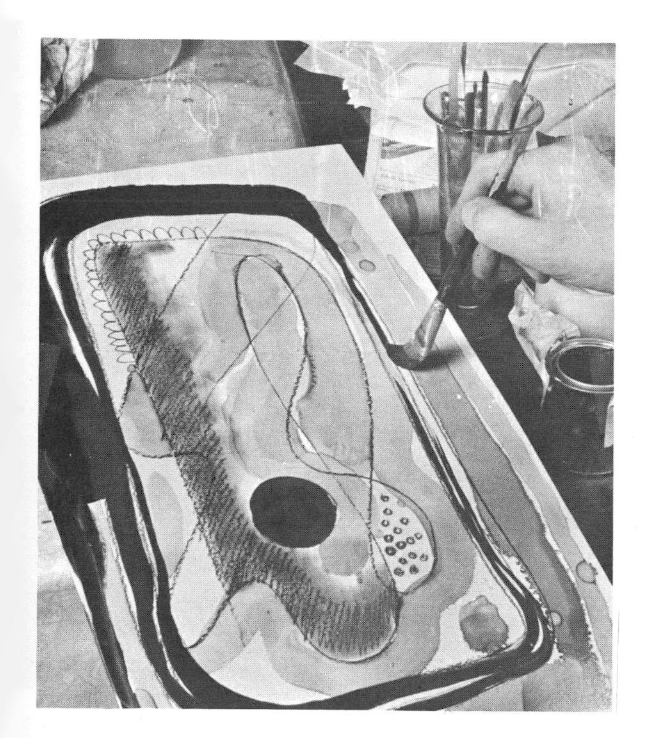

Oil-base enamel is used for dark areas.

Upward strokes show ease of application.

PLATE 12

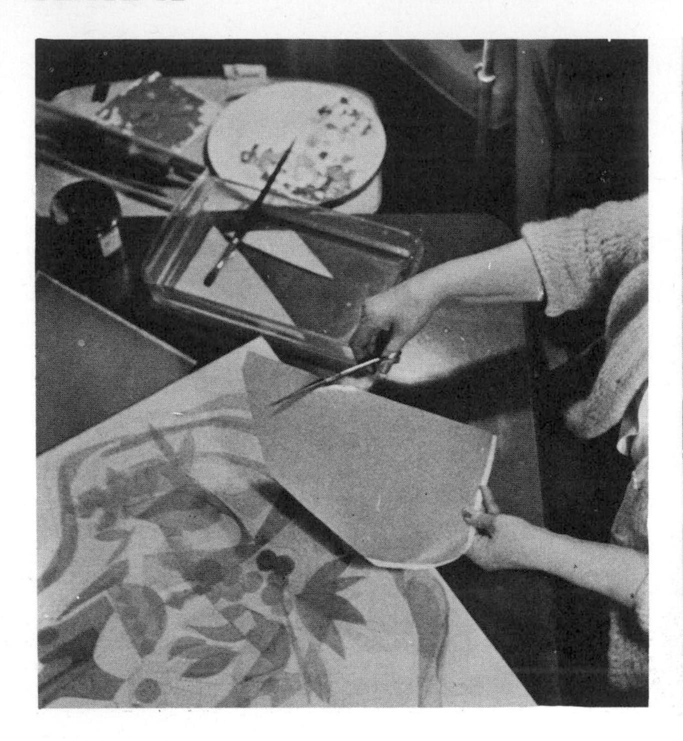

Glass enamel sheets (Decal type) in a wide range of colors are cut with scissors or sharp knife to make collage patterns.

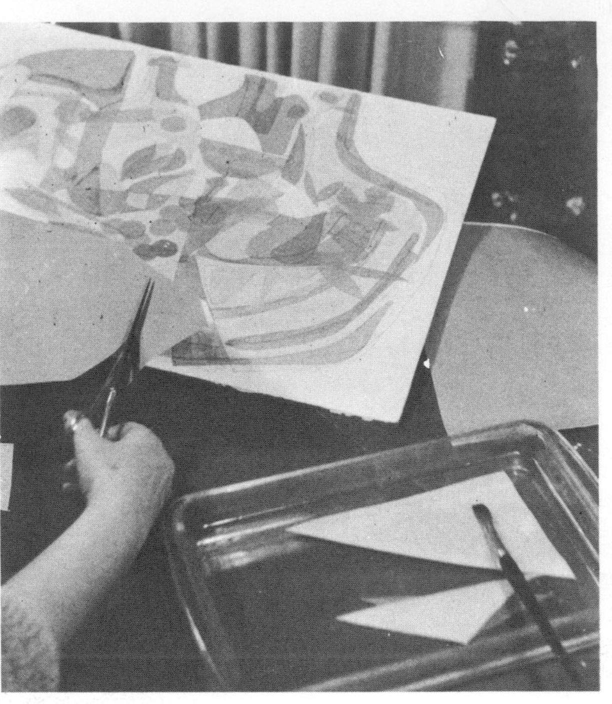

Cut pieces are placed in warm water which serves to remove the gelatine backing.

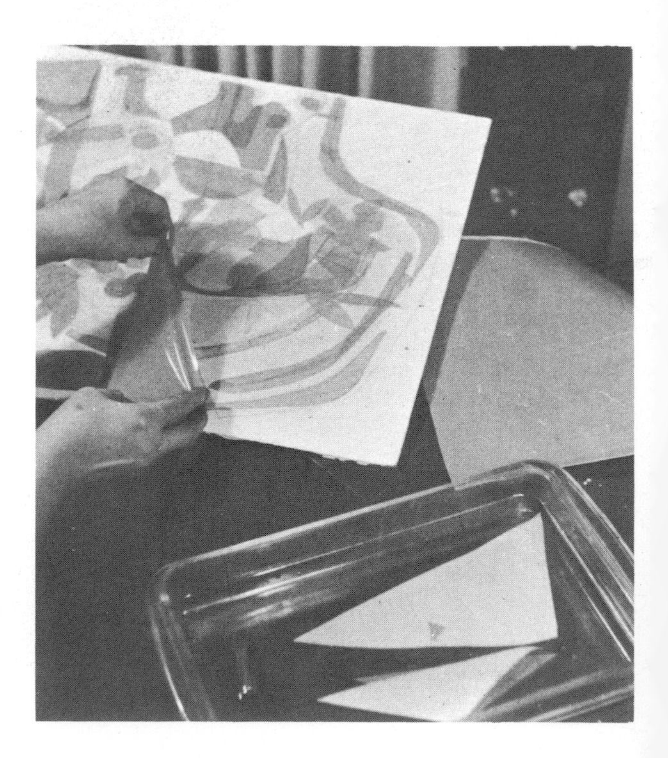

Sheeting of enamel is placed on the panel.

Painting executed over a beige ground.

Stylized panel is a most effective rendering. By Thelma Winter.

An intermixture of yellows, whites and browns produces warm earth tones.

Enamel colors are ideal for portrait painting.

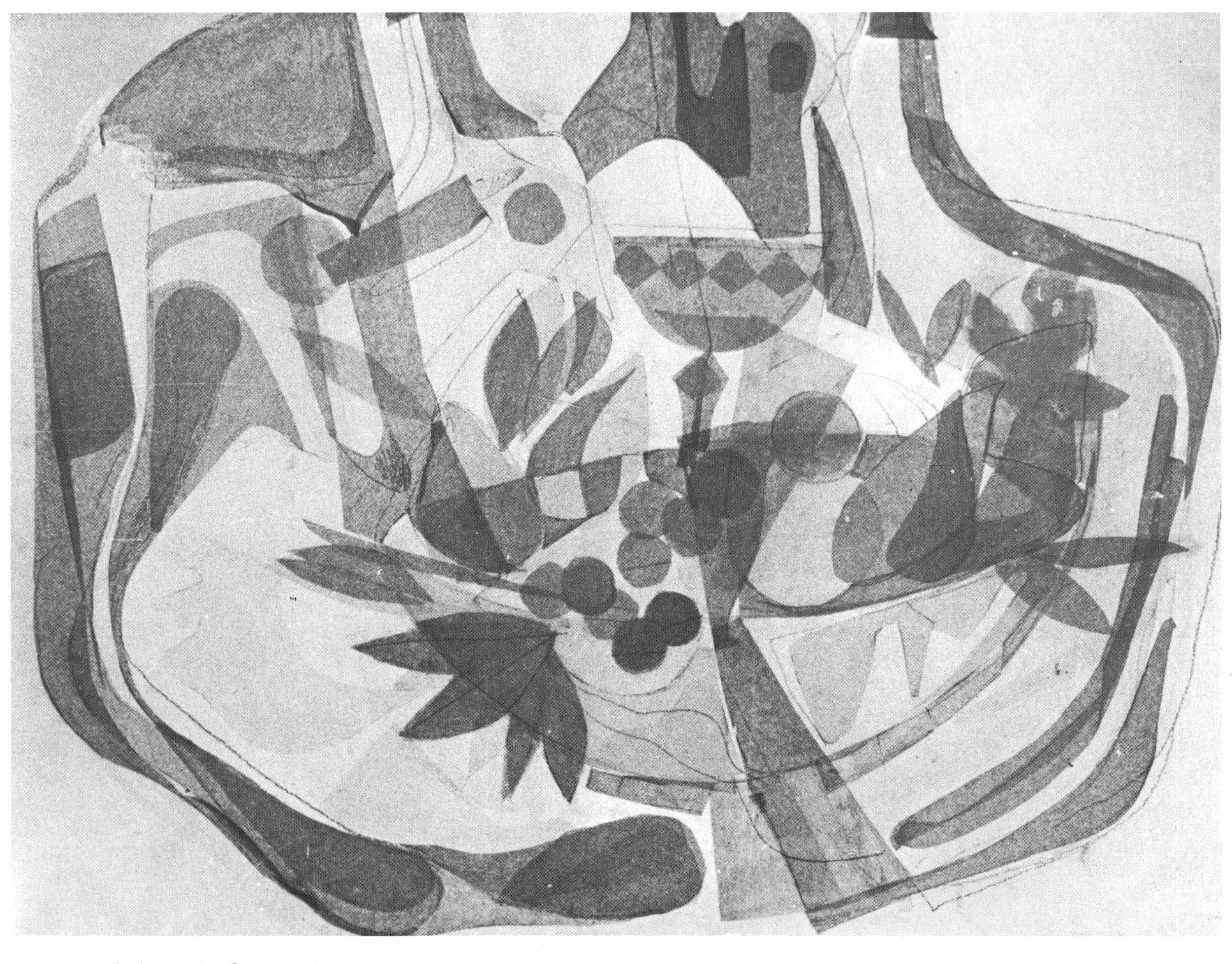

Application of color sheeting follows the black line drawing.

PLATE 16

Tiles are decorated with colorful enamel sheeting.

A blotter or absorbent paper is used to absorb moisture from under the sheeting.

Firing an individual tile at 1300°F for two minutes.

Colored sheets hang in the background. The tile has abstract decoration.

PLATE 18

Sunflowers in a vase are freely drawn over tiles with brush and brown enamel.

White enamel steel tile is easily removed from section.

Numbering the back of the tiles will aid in easy assembly after firing.

Form in the composition takes shape by adding yellow, dark brown and black.

Frames placed around the tile section to test compositional effect from a distance.

PLATE 20

A brush is used to tone areas of the landscape.

Distant hill laid in with palette knife.

If enamel paint is thinned properly a paint roller can be used in application to large areas.

Using paper as a straight-edge the paint roller can produce interesting effects.

PLATE 22

A bristle brush is used to compose still life.

The palette knife is used to lay in color.

Enamel steel panels showing variety of color, design, texture and gold.

Enamel plate designs by Thelma Frazier Winter.

Enamel on copper accessories by the author [1969].

Enamel painted plates, Flora and Fauna. By Thelma Frazier Winter [1958].

Decorative accessories in blue-green by the author [1969].

Transparent enamel on steel and copper. By the author [1969].

Liquid gold is applied with pointed brush.

A section of 'David' is painted with color.

PLATE 26

Textures can be obtained by use of pocket combs or serrated tools.

A wide assortment of camel's hair and bristle brushes are used for covering surface and detail work.

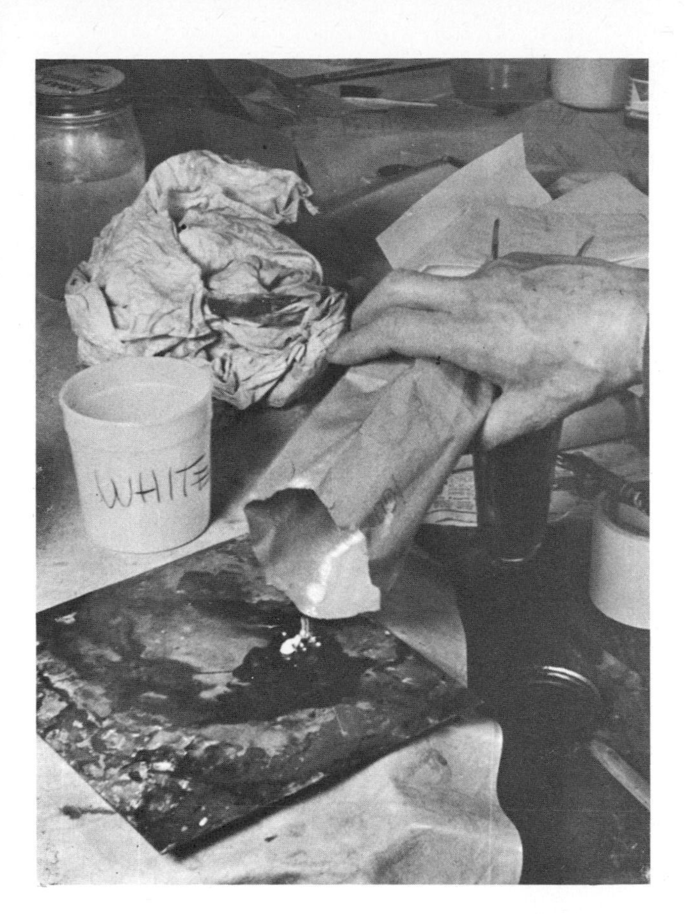

Enamel powders can be added to change color values or to thicken paste.

colors lend themselves to finger painting.

PLATE 28

Large areas can be covered quickly with a wide brush.

The sharp edge of razor blade will produce an unusual surface.

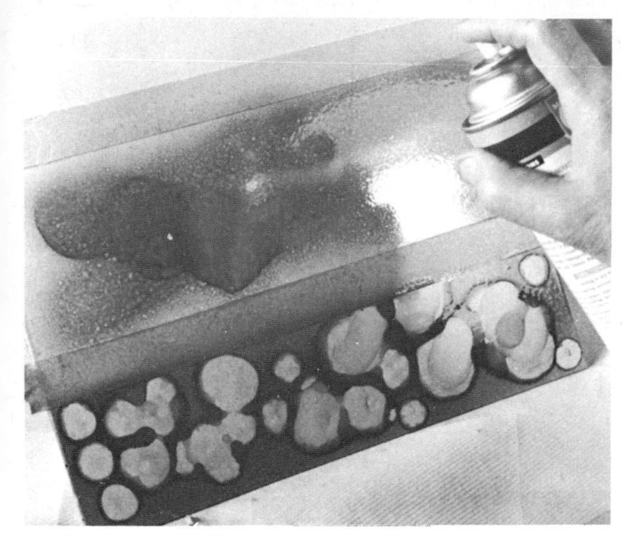

Masking tape is used to protect areas that will not receive the lustre spray.

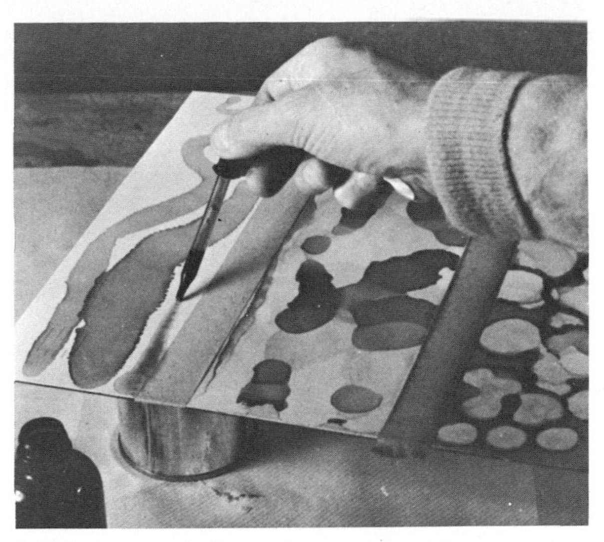

Gold lustre is applied to a white enamel panel with an eye-dropper.

A few drops of ethyl acetate applied with an eye-dropper will produce circular shell-shaped patterns.

Applying small drops of ethyl acetate to the gold lustre.

PLATE 30

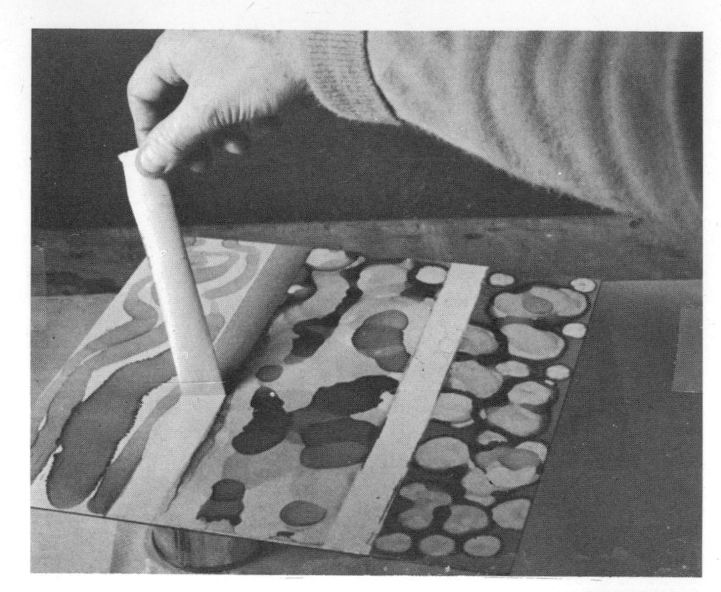

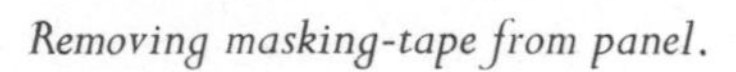

Removing masking-tape from panel.

String is laid on the white enamel panel.

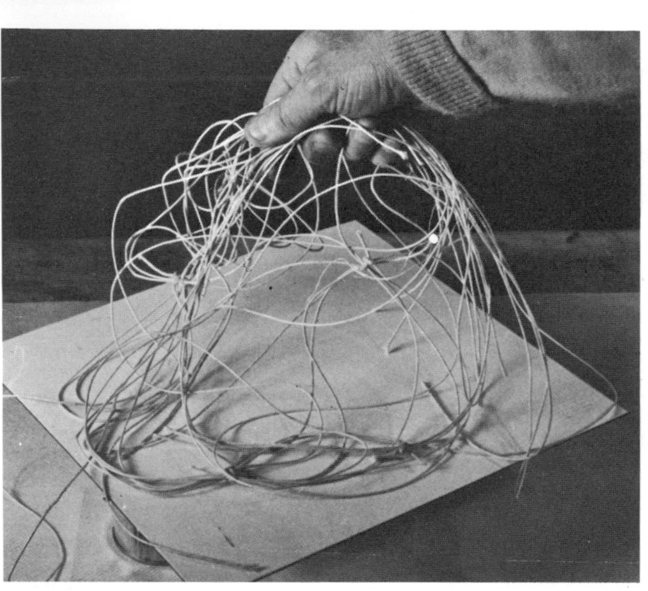

Liquid gold lustre sprayed over string with mouth blower.

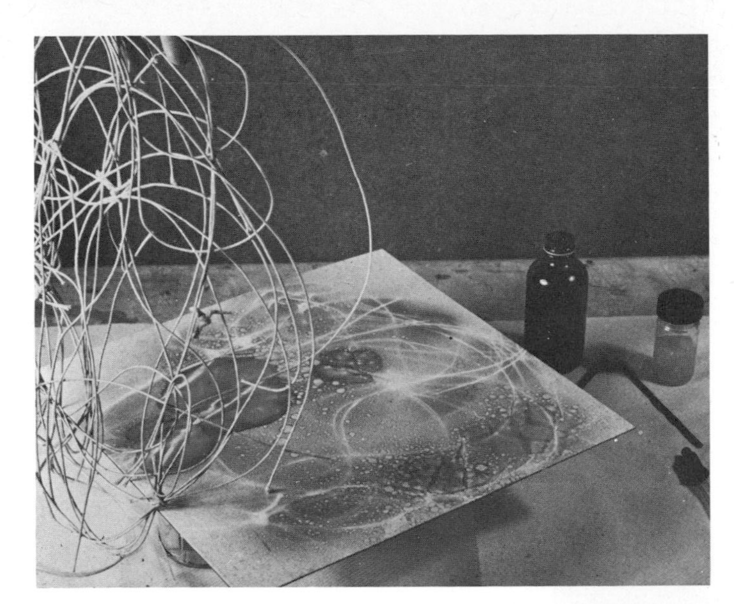

String is removed from the panel leaving line impressions.

The size of the circles will depend upon the amount of ethyl-acetate applied.

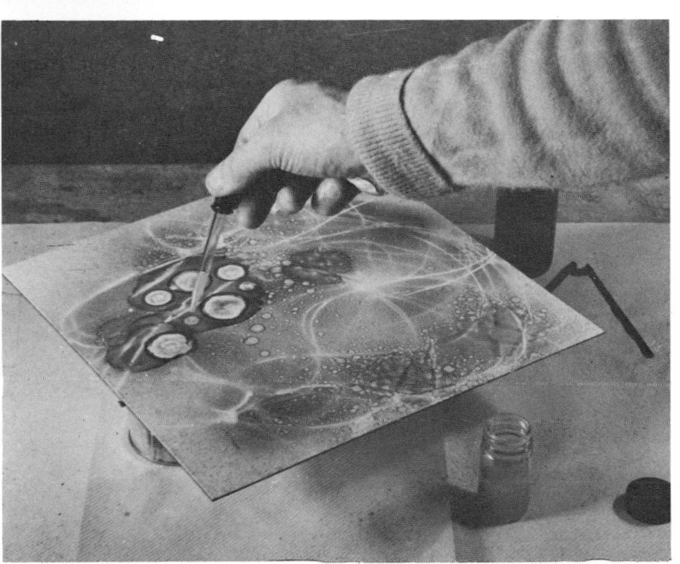

The panels were fired at 1 350°F for 3 min.

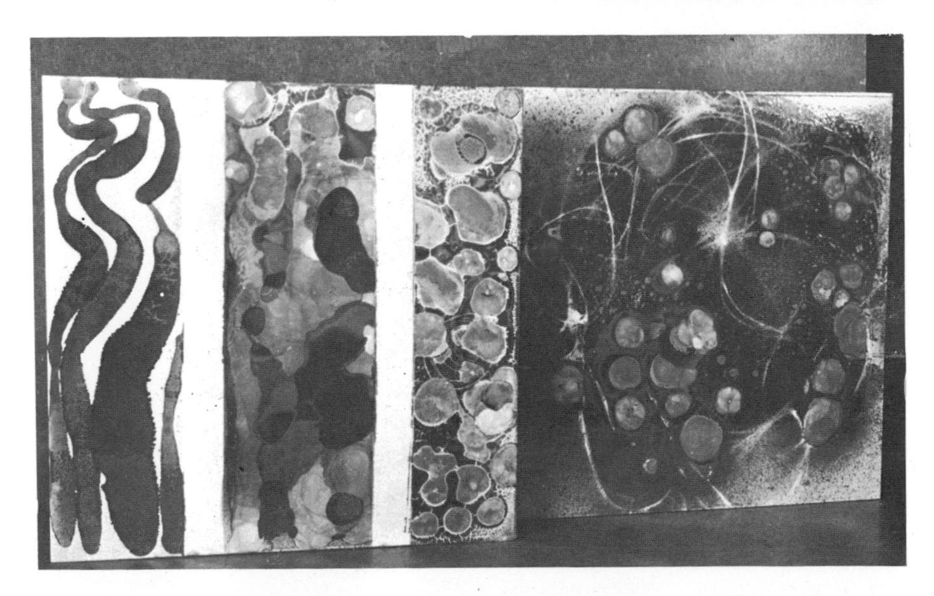

PLATE 32

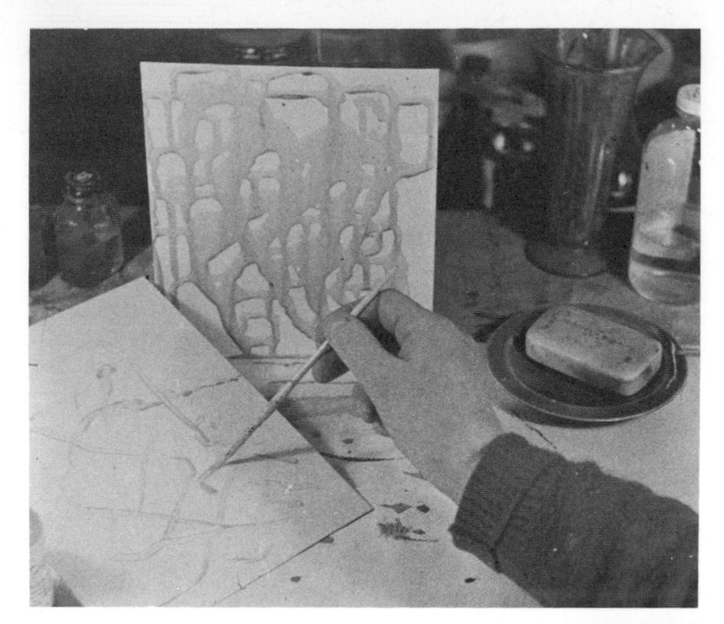

Liquid rubber maskoid is painted on a fired white enamel surface.

Liquid gold, lustre colors or enamels can be sprayed with air brush, gun or mouth blower.

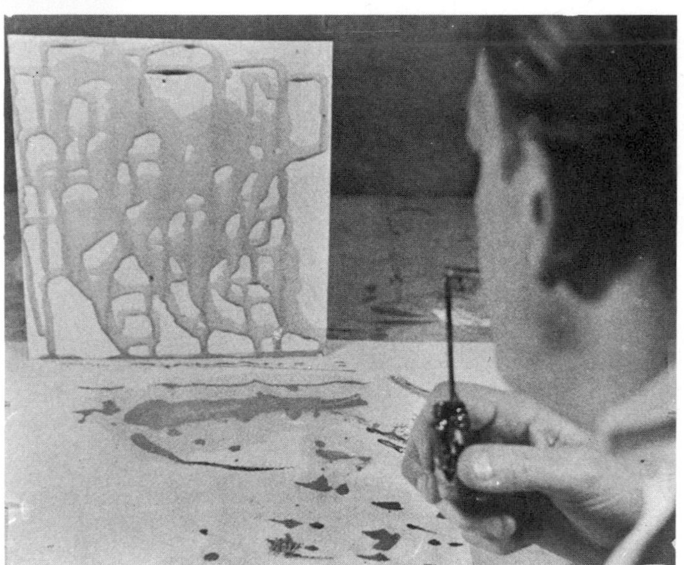

Rubber film can be peeled off revealing unusual pattern.

Bare copper pins must be coated with white or transparent flux before colors are applied.

A touch of color is applied to the horse pin.

Necklaces in a wide variety of colors—many with silver and gold paillons (foil).

PLATE 34

Other patterns and textures can be acquired with sharp or blunt tools, razor blades, combs and pointed sticks.

Doves, an enamel plate by Thelma Frazier Winter.

Crucibles can be used to melt enamel in making strings or lumps for texture. Pouring molten enamel into water will produce small particles, called frit.

Small round particles applied with fingers for textural effects.

PLATE 36

A unique surface made from lump and string textures.

Texture surfaces combined with liquid enamels have variegated surface qualities.

Black line drawing is now possible with graphite pencil on a white or colored matte enamel surface.

Shading is done with the flat side of the pencil.

The finger can be used to produce tonal effect.

PLATE 38

A paper stomp can also be used to tone in an area of the drawing.

Graphite pencil, white pencil and stomp can be used effectively.

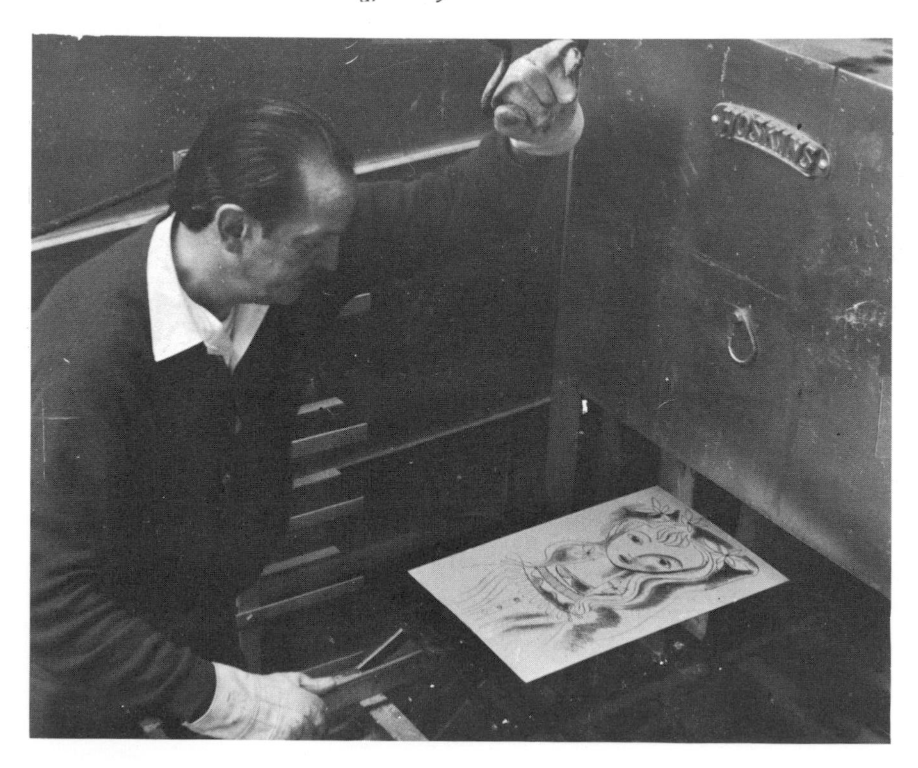

Drawing being fired at $1\,400^{\circ}$*F for* $2\frac{1}{2}$ *min.*

White line drawings are possible on blue-black ground coat surfaces or any dark enamel surface that is not glossy.

PLATE 40

Black line drawing is effective over white matte surfaced tiles.

A chisel-edge pencil produces both wide and narrow line drawings.

Regular colored ceramic pencils work well ov such surfaces.

Liquid enamel colors can be applied with pointed brush.

Dark tonal areas can be made with pencil or dark color.

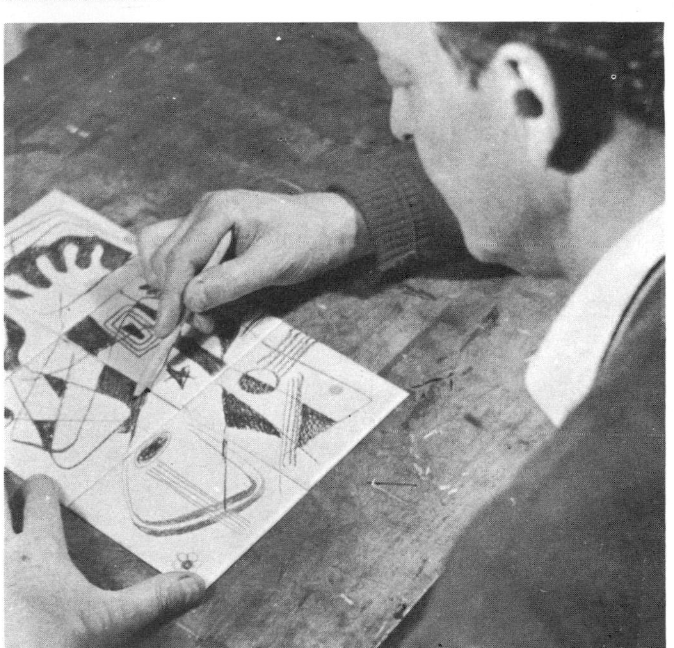

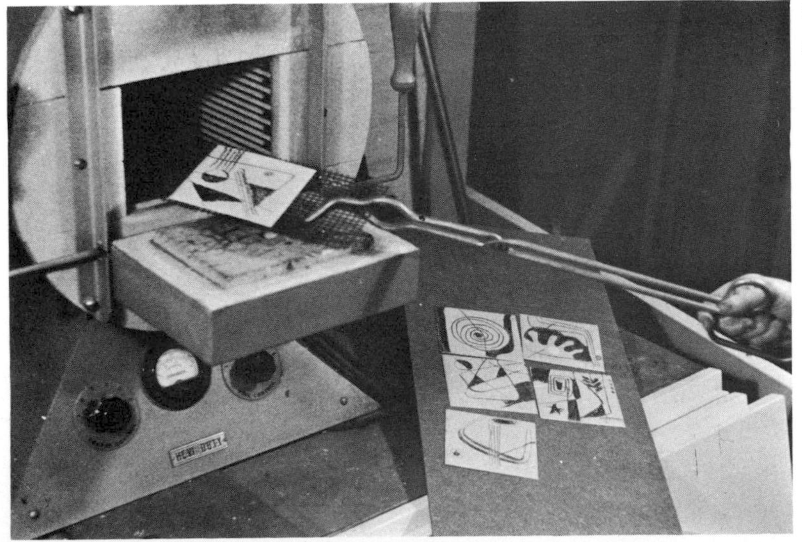

Tiles are fired individually for 2½ min. at 1 400°F.

PLATE 42

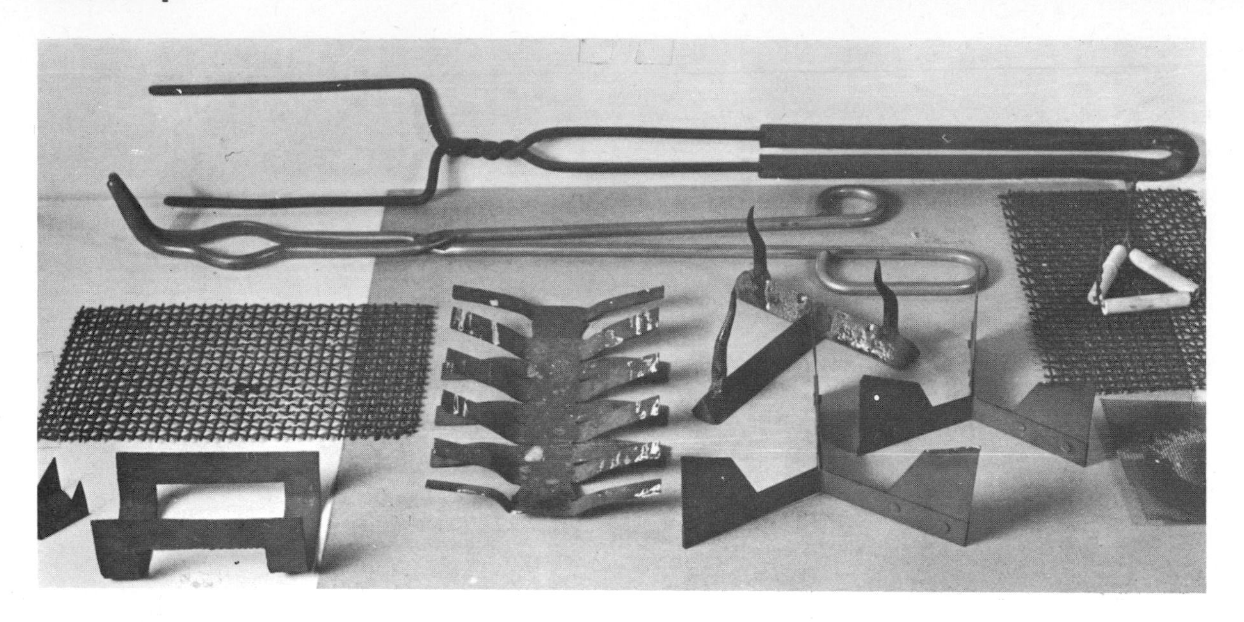

Chromel wire screens, planch, tripods, tongs and fork are used to hold objects while firing.

White drawing pencil of special ingredient is used to draw over dark-tone matte surface tiles. Lines will not fade in firing.

Liquid gold or enamel colors can be painted in areas of the tile.

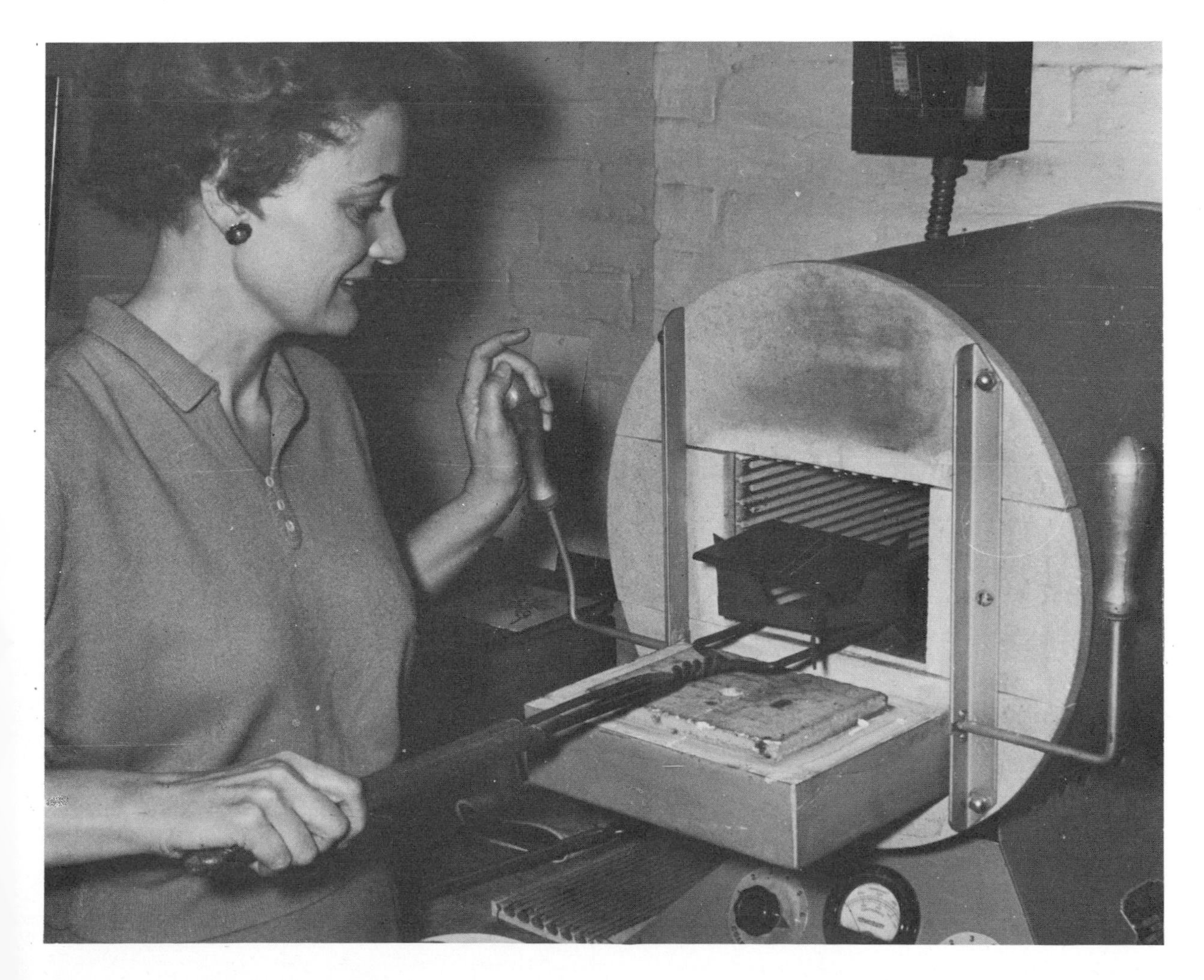

White line drawing is fired at 1 400°F for 3 min.

Effective linear drawing is made with white pencil.

Tiles, ash tray and lighter insert can be made using white line technique.

8 Working with the Palette Knife

THE artist can open up surprisingly new effects by allowing his intuition and intellect to work for him. New creative regions swing into use and delightful affects appear spontaneously. A few cunning strokes of a springy palette knife or the side of the finger running over a wet surface of enamel can bring out imaginative and realistic images of great strength. Many persons are afraid of displaying their true emotions and letting themselves go; this hides their intuitive power. While the brush in its variety of styles is the artist's all-purpose tool, the palette knife is capable of a wide range of specialized techniques. The knife can make clearly defined lines and chisel-edge planes; it can be used to apply enamel in wide dashing strokes, heavy at one end and thin at the other. Brilliant patterns of light and dark values can be built up with an impasto effect, and small dots of paint can be applied with the tip of the knife. A multiplicity of textures can also be built up by using assorted color patterns.

Through practice the artist can experiment so as to achieve the appropriate depths for successful fusing. He must guard against applying the enamel too thickly so that undesirable crackling effects can be avoided. Such effects may make the enamel tear off the metal.

Portraits, landscapes, still life, abstractions, and many other types of painting can be executed with the palette knife. Using it as a creative tool you can do more with it than merely lay on color or cover a large area of a panel. It is a tricky tool but one with which you can have much fun.

Abstract Expressionist and Floating Colors

LIQUID enamels and metallic lustres have a life of their own, whether used in semi-inert substance or as wholly transparent and opaque glasses flooded on to a surface with water or oils and turpentine. The images they create grow on the panel like flowers or other moving forms. Such paintings are dramas of ebb and flow and of tensions that may sometimes explode off the plane with sudden violence. The descriptions of the effects that can be obtained are as exciting as the enamel paintings themselves: cataclysmic, molten lava, hurtling debris, or cosmic storms.

Effects simulating marble finishes are produced by spooning on liquid enamels and floating an assortment of variegated colors on the base liquid. With drops of water or turpentine (depending on which type of enamel is in use) one can set the enamel in motion to form patterns. Small hair-lines of color can be painted with a pointed brush or a feather, and a coarse brush or comb can be used to pull and get the colors moving. Experimenting with less viscous media such as alcohol, ethyl acetate, and benzene will also prove exciting. In using these materials one must remember to smoke them before placing them in the intense heat to fuse. This 'smoking' process, which simply means removing the volatile materials from the drying enamels, can be done by leaving the articles near the door of the kiln for an hour or so. The purpose of this is to prevent boiling and movement in the enamel coats which could cause faulty articles.

9 Sparkling Metallics

IN many stores, and especially hardware stores, one finds small bottles of sparkling metallic paints on display. They come in silver, gold, bronze, copper and assorted colors. These colors do not always persist when heat is applied to them, but it is known that most of these paints are made of metallic copper. When heated the compounds responsible for the sparkling colors evaporate. These cheap paints meet the demands of the artist in his search for another luminous textural material. The minute particles of copper obtained by heating these paints can be sprinkled freely on to an enamel surface (fired or unfired) or placed in a design pattern. When the piece is fired at the usual 1 450°F for 2 or 3 minutes the intense heat will kill any remaining color and leave the particles securely embedded in the fired enamel surface. The heat will leave the metal dark and muddy looking but a metallic brightness can be recovered by polishing with metal polish or by using a brush, water and scouring powder.

The imaginative enamellist who stays alert and keeps his eyes open for new materials which produce unusual effects in firing can add richness and excitement to his favourite medium.

10 Enamel Costume Jewellery

SOFT, pliable sheet copper in 20–22 gauge is so easy to handle, using simple tools and the fingers, that a very wide range of whimsical and charming brooches, pins, and necklaces can be made. We are indebted to Harold Tishler, Long Island enamellist, for showing us how to make third dimensional flowers, cats, horses, rabbits, birds, butterflies, mermaids and seahorses, as well as the round, oblong and oval brooches made from enamel and silver foil in the form of paillons.

Harold Tishler is a meticulous craftsman who loves the medium. His many creations admirably demonstrate this. He and the author studied together in 1931–2 at the Kunstgewerbeschüle, Vienna, Austria. He has exhibited widely in the USA, and in 1937 exhibited his large enamels at the International Exposition in Paris, where he was awarded a gold and silver medal. He introduced a course of enamelling during 1933–5 at the New School in New York, and from 1950 to 1959 he managed a craft and sales shop at Rockport, Massachusetts. On a visit to Cleveland in 1935 I showed him the dipping process whereby metal could be immersed into liquid enamel to achieve a better and more complete coating than could be obtained from simply sifting on the powdered enamel. The fine examples of his work are proof of this effective and easy process.

Jewellery as a Craft

THE adornment of mankind with precious stones, metals and jewel enamels had its origin in the earliest days of history. Fired colored enamels became substitutes for precious gems, but it was not simply a case of imitation, for the art of enamelling has become great in its own right. The massive crown jewels of State and the sacred ornaments of religion, the wedding band and the ubiquitous buckles, pins and buttons for clothing, are but a few of the forms that have been wedded to the jeweller's art and craftsmanship.

Making enamelled jewellery today is an exciting activity for the amateur, hobbyist, student and professional craftsman. While we do not expect superior technical skills from the high school student and the amateur, we do believe that some of the simple forms of jewellery making and firing enamels on copper and steel will prove so fascinating that the newcomer will want to pursue the craft further.

One can readily understand that in enamel jewellery, design becomes a requisite. A knowledge of small space breaking, color pattern, and designing to any given shape is important before one commences with the techniques of enamel application and firing. Along with the design of an article, one must remember that jewellery when worn as an ornament must integrate itself with the wearer's costume as a whole, and must of course be complimentary and flattering to the wearer.

The Artist as Jeweller

THIS book does not aim at covering the complete field of jewellery making. It is hoped that the many examples illustrated will serve to stimulate and excite the reader into trying out new art forms. Small pieces of copper or steel in all shapes and sizes can easily be bent, twisted and hammered into imaginative objects. By using the many liquid glassy enamels available on the markets, any metal shape can be easily coated by dipping the object into the glass solution, drying and firing. One can ensure by careful application that all parts and edges of the metal will be coated, and this coating is then receptive to further coatings of transparent colors, textures or designs.

It is customary to think of the jeweller as a specialized craftsman, dealing in glittering diamonds, brightly colored gems and precious metals for a wealthy and discerning public. Today the jeweller comes in for competition from the painter and sculptor who are making jewellery as a means of personal expression as well as for the commercial market.

Sculptors now working in this field think of their small works as being witty and playful, sometimes simple, and in some instances severe. The unpredictable flow of molten metals or the flow of jewellery enamels on red hot metal in a kiln can often be subconscious preparation for larger works. Others might experience the opposite feeling, that jewellery is a way of fully expressing the accidental and of freeing oneself of decorative elements one does not wish to use in sculpture.

Both the painter and sculptor are alert to the technical achievements and discoveries of our industrial laboratories. Materials which nature has given us are being improved and changed so that their use is spreading rapidly. The creative artist, painter, sculptor and designer are keenly aware of these developments in art media, and are bringing a fresh, experimental approach to traditional art forms.

Findings

THE term *findings* is used to designate the working mechanisms on jewellery which are usually purchased as prefabricated units from commercial suppliers. Supplying jewellers and craftsmen with these engineered parts has become big business, and many erstwhile small supply houses have grown into big corporations. Some determined professional craftsmen take great pride in making their own fittings and attachment devices, but for the price conscious patron or consumer in the retail market time becomes a most precious commodity and the craftsman need not be ashamed to take advantage of precision-made parts sold by reliable engineering suppliers.

For brooches and other types of pins, fastening wires can be soldered either with hard silver solder which fuses at about 1 450°F or with soft solders which can be formulated to fuse at the temperatures of a lighted match or the stroke of a torch flame within a few seconds. Soft solder findings usually have a larger base to provide better contact with the metal to which they are soldered. Soft solder is usually made of equal parts by weight of lead and tin, and melts at about 370°F. Another type of easy flowing hard solder can be purchased in round wire form with an acid core acting as a flux. Hard silver solder is supplied in small sheet form which is then cut with tinning shears into small bits and laid on the edges of the metal to be soldered. Both metal parts to be soldered should be previously cleaned with acid or fine steel wool. Powdered white borax performs well as a flux when hard silver solder is in use.

Findings are produced in brass, sterling silver, gold filled, and carat gold. Suppliers' catalogues show illustrations of the following:

- Earclips
- Earwires (screw type)
- Cufflinks
- Tie clasps
- Spring rings
- Foldover catches
- Chains
- Eye pins
- Jump rings
- Hinges
- Foldover clasps
- Pinbacks (brooches)
- Keyrings
- Metal mountings
- Sister hooks
- Necklace and bracelet clasps
- Joints, catches and pin stems
- Button eyes
- Frame settings
- Adjustable rings
- Money clips
- Pendants

These are just a small number of the many valuable aids for the craftsman.

If findings are hard soldered to the back of brooches already enamelled, or pieces of metal to be enamelled, small bits of clay can be applied to the soldered spot to prevent sticking when the enamel is placed in the kiln for firing. Skill comes with experience and the simple rules for good jewellery enamelling are soon learned. The 'learn by doing' method is best for personal growth, while precision work demands meticulous handling of tools and materials; the person with the best attitude will be most at home with techniques that demand skill and thoughtful care.

11 Sculptural Forms Made With Shears and Pliers

THE extremely fine cutting of animals, birds and butterflies mentioned in the previous sections is executed with a fine metal coping or bandsaw (it is usual to cut a small pile of eight or ten simultaneously). The shears and pliers are used to cut the heavier metal blanks and to bend and twist the various parts of each object.

Shears and pliers are excellent tools for the ambitious sculptor interested in larger constructions, such as free-form shapes, stabiles, mobiles and kinetic sculptures. A good range of metal stock from which to choose is available. It includes various gauges of wire, rod, tubing, sheeting, perforated strips, embossed sheets and cast metal forms in both copper and steel. Several informative books have been published on the subject of metal sculpture. Unfortunately none of these books shows the process of coating metal forms with glass enamel for added color, texture and permanence. Atmospheric conditions will rust and corrode metals treated with common paint or acrylic coatings, whereas fired glass surfaces will withstand all forms of atmospheric and chemical attack.

Copper

COPPER is the metal perhaps most commonly used by the enameller. Much softer than steel, being ductile and malleable, it behaves admirably in work which requires stretching by hammering and bending. Transparent enamels can be fired on to it directly to produce colors of depth and luminosity. This richness results from the sheen of the polished copper surface that is visible through the fired enamel. When 18 gauge copper is hammered and planished, virtually every hammer mark will give off numerous reflected lights and color effects.

Copper is most commonly available in sheet or roll form. The sheets measure 30 in × 60 in or 20 in × 90 in, and rolls may be bought in 6, 8, 10, 12, 14 and 16 in widths (10 in = 254 mm). Copper can also be bought from the supplier in various size tubes suitable for making tall vases. One metal manufacturer also sells small sheets of copper with unusual embossed designs such as radiating lines, stars, and cross-hatching, many of which are machine engraved or die stamped.

Only pure copper can be enamelled successfully, so it is important when ordering to ask for electrolytic, cold-rolled, annealed stock. Never buy roofing copper which is an alloy containing zinc and other metals. Once this metal is subjected to the heat required in enamelling, cracks and blisters will appear in the enamel during the second firing. Since most enamelled pieces before being completed must be fired three or four times it is obvious that this impure metal is not suitable for high-class enamelling.

The question of enamelling brass or bronze often arises among craftsmen. These metals are also alloys. One-third of the weight of brass consists of zinc, and this makes it very difficult to enamel, because upon cooling the enamel will pop off. Bronze, which is composed of 90% copper and 10% tin, is unpredictable and difficult to work, but on a heavy cast form it can be used for nameplates and monumental tablets, providing a low-temperature opaque enamel is used.

However, there is one alloy, a wonderful reddish copper called *Tombac* which our enamel class used when I was a student in Vienna. It was a specially smelted copper containing a small amount of zinc. Although we do not have this particular type of metal in the USA, the alloy which comes closest to it, I have discovered, is *guilders' metal*. This contains 95% copper and 5% zinc. It can be used for firing transparent ruby red enamels directly on to the surface without first applying an undercoat of flux to the metal.

Before ruby red is fired over ordinary copper, a transparent flux must first be used for successful results. This will be considered further in the section dealing with the enamelling of copper.

The thickness of copper is determined with the Brown and Sharp gauge which is a system of numbers ranging from 3/0 to 36, each indicating a certain thickness in thousandths of an inch. The larger the gauge number, the thinner the metal. A 'number 14 gauge' sheet is about one-sixteenth of an inch thick. One should work with at least an 18 gauge copper since this heavy weight produces the best work. For most high-grade enamel pieces several firings are usually necessary to complete the piece; heavy copper will best stand up to numerous firings. The thinner gauge copper, such as 20–22 or 24 can of course be enamelled successfully but it requires greater skill and experience to obtain the desired results. There are many commercial die-stamped and spun copper shapes on the market. Most of them are unsatisfactory as regards proportion, shape and design, but their greatest drawback is that they are produced from copper that is too thin for successful enamelling. The intense heat of the furnace is too much for the thin metal and several successive firings will be injurious to the piece. Objects made from heavy copper have the feeling of weight and quality which greatly adds to their value and appreciation by the owner.

Craft classes in many schools have been working with copper for a long time and students have learned the ease and fun of working with it. Now they can experience the added thrill of applying and firing vitreous enamel to it.

ENAMELLING COPPER

Copper can be successfully enamelled in several ways:

1. By dry process application through a sieve.
2. By dipping the metal into a basin of liquid enamel slurry or slip.
3. By spraying.

Consideration must be given to the choice of the method since the final surface of the objects will vary in each of these techniques.

Traditional enamellers used a steel spatula or pointer to apply the ground enamel in wet form (grain by grain) to the copper. This method was painstaking and produced stilted results. Large surfaces of metal could never be covered in this manner. In the course of teaching the first class of enamelling to be offered in the USA, I was able to show my students an easier and faster method, which involved sifting the 80 mesh enamel powder on to the metal by means of a large sieve. The sieve, which is about 5 in diameter, can be easily held and manipulated with the hand. The screen opening of 80 mesh per inch allows the 80 mesh enamel to fall through with a gentle shaking of the wrist. Both opaque and transparent enamels can be applied to copper by the sifting method, but the most highly transparent enamel colored surfaces can be obtained only by the dry application technique. With this method the surface of the copper must first be coated with a thin solution of gum tragacanth, which acts as a light adhesive to hold the powdered enamel on to the metal until it is fired. Enamel suppliers sell transparent enamels milled in alcohol for dipping or spraying, but the colors when fired are rather thin and washy. Only the dry process yields deep jewel-like textures and high-quality surfaces.

Copper can also be treated with opaque enamels by dipping or spraying. Commercial suppliers sell all colors as well as black and white. This enamel is prepared by grinding clay and various deflocculants (inorganic salts) together in a porcelain ball mill for 3 to 4 hours. The inorganic salts act as suspending agents and deflocculants which keep the slush or slip enamel in a creamy state. The enamel used for dipping and spraying must be ground fine enough to pass through a 200 or 250 mesh screen.

Accompanying photographs illustrating the dipping of copper animals, flowers, and other forms into liquid enamel slips show graphically the ease and advantage of this method.

The use of the airbrush is another easy technique for applying enamel to copper. The extremely fine enamel must be thinned with water if it is to pass through the finer guns. Suppliers will suggest the proper equipment.

If larger spray equipment is needed the common compressors and air guns available are used. They require at least 35 to 40 psi air pressure.

Copper is my favourite metal, for when handled with care and respect it responds accordingly. By using transparent enamels on copper one can produce bowls, plaques and murals of great quality and technical excellence. From a record of my enamel works exhibited in museums throughout the world I find that the majority were executed on copper.

Many people, both young and old, have a tendency to shift from one metal to another because they want to acquire the tricks and unusual effects before they have developed technical mastery or skill in working with copper. Each metal that can be enamelled has its own distinct characteristics, and a craftsman could easily devote a lifetime to any one without exhausting its decorative and utilitarian potentials.

12 Steel: How Man Learned to Use the Metal from Heaven

THE abundance and versatility of steel have made this metal so common that it is regarded as just another basic commodity, like salt or sugar. In fact the manufacture of steel by modern methods began scarcely a century ago.

Steel is produced by resmelting pig iron to which carbon is added to transform it into carbon steel. Other elements are often included to give it special properties. The use of iron by man can be traced as far back as 3500 BC.

Early Egyptians called it 'the metal from heaven', since it came from meteorites. The words applied to iron by Assyrians, Babylonians, Chaldeans and Hebrews meant the same thing.

Though it is one of the most plentiful elements on the Earth, iron did not come into general use for another 2 000 years.

Iron comprises about 5% of the Earth's crust. Unlike other metals, such as copper, silver and gold, it is hardly ever found free in the earth. Instead it is locked tightly in chemical combination with oxygen, sulphur, silicon, and other impurities. Some iron-bearing compounds are also found with clay, sand and gravel, or even embedded in solid rock.

For at least 5 000 years man has been working to free this elusive and valuable raw material from the Earth's crust.

Ancient man stumbled on a discovery. If he heated certain kinds of black and reddish earth in a very hot wood fire he could extract small lumps of spongy stuff which could be hammered into daggers, knives and other useful implements.

Later artisans learned to build primitive forges to convert the iron ore into metal. The oldest forges were crude hearths on which the ore was piled and partly melted by the heat from burning wood or charcoal. Draughts of air were later applied to speed combustion and liberate the ore's impurities. From this basic idea were developed the crude bellows used to blow the fires of ancient forges. In time clever craftsmen found that by giving iron a special treatment they could turn it into the stronger, less brittle, more valuable metal, now known as steel.

Through the centuries the art of making and fabricating iron, and even a little steel, was developed in a number of civilizations in ancient Europe. Almost always it was a jealously guarded secret which brought its possessors wealth and power.

The Moors who invaded Spain carried with them their steel-making knowledge, and the city of Toledo became famous for its fine steel blades. At the other end of the Mediterranean craftsmen were equally celebrated for the keen edges of the swords they fashioned from a special kind of steel. Among the early settlers in America were men who had a knowledge of iron making.

Iron ore was discovered on Roanoke Island in 1584 by an English expedition sent to the Carolinas by Sir Walter Raleigh. The Virginia colony sent ore to England in 1608 and eleven years later a number of skilled iron workers arrived to build a forge at Falling Creek. This was destroyed by Indians during the massacre of the entire colony in 1622.

The next important venture in iron making was in the colony of Massachusetts where iron ore was discovered in 1629 along the Saugus River. In 1642 samples of the ore were sent to England to arouse interest in forming a company to exploit the ore. Five thousand dollars was raised and the workmen and raw materials were despatched to build a foundry near Lynn. On a level with the most advanced contemporary furnaces in England it became known as the birthplace of the American iron and steel industry.

The second New England ironworks was established at Braintree, south of Boston, in 1646, and the third at Taunton in 1652, which remained in operation until 1865. Demand for anchors, nails, bolts, chains, and other gear caused many more forges to spring up in eastern Massachusetts. The discovery of iron in the limestone valleys to the west soon drew people westward, and by 1750 there were scores of iron works in operation in this region.

Enamels and Formulas for Sheet Steel

THERE may be a few artists with a technical mind who will want to learn how enamels for steel are made so that they can make their own glass melts. Even if this is not the desire of the majority, technical information on this subject is most valuable.

Four types of glasses are used for sheet steel: clear, zirconium, antimony, and titanium. The type of glass frit used and the color selections are interdependent. The strongest and brightest colors are obtained in the clear glass. Moderately strong to bright colors can be produced with semi-opaque, and moderately strong colors with titanium glasses. Technicians have found that the most stable milled-in colors come in the range of ivory-yellow, yellow, yellow-brown and blue-green. Shades of pink are easily produced in most glasses, but the crimson to purple color range must be smelted-in, using gold salts or manganese oxides.

Research has shown how each oxide behaves in all types of glass frit. Each oxide has been rated for strength, purity, and stability, and the results show that only a few base oxides performed well in all glasses. Most oxides were acceptable in one or two types but some were unsatisfactory. A chart was compiled to show the color frit performance rating. The glasses considered were as follows:

CLEAR GLASSES

These produce a good, dark, strong intense color in blue, red, yellow and black. When cadmium reds or cadmium yellows are used the glass has to be cadmium stabilized.

ZIRCONIUM GLASSES

Most color oxides are compatible with this type of glass but the cadmium family is a major exception and is not used. Though zirconium glass was at one time very popular in industry it has given way to titanium frit because of the latter's greater opacity and higher acid resistance.

ANTIMONY GLASS

This type which, although it has served nobly for many years in many color formulations, is now being replaced by the more opaque titanium-bearing frit which offers better physical and chemical properties.

TITANIUM GLASS

A glass available in opaque and semi-opaque form. The variation in opacity is controlled by the recrystallization of the titanium. This glass is well suited for light pastel colors, whereas the semi-opaque variety is used for darker colors.

In many instances this type of color accuracy and control is quite outside the needs of the artist painter but it is useful to know how color accuracy can be maintained. Often the artist is called upon to match the colors of his glass with other materials such as woods, metals, plastics, fabrics and paint.

In due time this art may grow to the point where artists feel the urge to experiment with or are compelled to smelt and produce their own colors in small pot melts. In this case the above technical information will prove useful.

A formula for a blue ground coat sheet iron glass:

Feldspar	29·20
Silica sand	20·50
Borax	29·30
Soda ash	8·80
Soda nitrate	4·90
Fluorspar	3·90
Cryolite	2·00
Manganese oxide	0·60
Cobalt oxide	0·40
Nickel	0·40
	100·00

A formula for a titanium white cover coat:

Dehydrated borax	20·00
Silica	37·00
Feldspar	8·90
Sodium nitrate	8·90
Potassium carbonate	1·70
Titanium dioxide	18·00
Mono ammonium phosphate	1·10
Potassium silicofluoride	4·40
	100·00

A formula for a dark blue glass cover coat:

Borax	27·80
Feldspar	25·80
Silica	22·30
Cryolite	4·30
Fluorspar	5·30
Soda ash	7·30
Soda nitrate	4·30
Manganese dioxide	1·00
Cobalt oxide	1·40
Iron oxide	0·50
	100·00

Formula for producing a clear glass for sheet iron:

Borax	27·80
Feldspar	25·80
Silica	22·30
Cryolite	4·30
Fluorspar	5·30
Soda ash	7·30
Soda nitrate	4·30
	97·10

Preparing Glass Enamel for Spraying and Dipping

WHEN it is decided which of the four frits is to be used, bearing in mind color and tone intensity needed, a batch of glass (a 1 500 gramme mill will yield about a quart of liquid enamel) is ground in a ball mill with other ingredients, including bentonite, potassium carbonate, clay and water. These are included so that the slip enamel will be of a creamy consistency and will perform well in spraying and dipping. The above mill will take about two-and-a-half hours grinding to produce a fine slip that will pass through a 200 mesh screen.

CLAY

White burning-clay (1 to 8 parts per 100 parts of frit) used in the mill serves two purposes: it acts as a suspending agent (that is, it keeps the small particles of glass from settling); and it produces green strength, making it possible for pieces of glass-sprayed enamel to be handled without causing any enamel to drop off. The clay particles also serve to bond the enamel to the metal being coated. Clays should contain the minimum of iron oxide, gypsum and other similar impurities. Kaolins and white-burning ball clays can be used. Using too much clay in a mill will make the glass harder to fuse and produce a greater degree of mattness.

A normal charge for a wet process mill is as follows: 100 parts quenched glass (frit), 6 parts clay, $\frac{1}{4}$ part potassium carbonate, and 40 parts water.

Dipping sheet iron in glass slip is faster than spraying and if the metal shape is complex, with undercuts and third dimensional planes, the piece can be held at the edges with tongs or the fingers and immersed in a vat of the enamel slip. After giving the piece a few twists with the wrist in order to drain off any excess slip the piece is placed upon a nailboard for drying. When completely dry it is ready for firing.

A normal type of charge for acid resisting dark colors is as follows:

100 lb clear frit,
5 lb clay,
$\frac{1}{2}$ lb potassium carbonate,
38 lb water,
and color oxide as required.

ENAMELLING SHEET STEEL

Traditional craftsmen used gold as a base metal upon which to fuse delicate glass colors. In due time other metals, such as silver and copper, were used. With successive firings of opaque white enamel on copper it was found that the white was usually discolored with a greenish cast from the copper oxide burning through the enamel. While certain grades of iron made in Germany in the late 1800s were being successfully coated with glass for utilitarian purposes, it was not until 1909 that an American company first offered a low-carbon, high-grade, cold-rolled steel as a base for glass enamels. These coated metals appeared on the market in the form of stoves, cooking utensils, sinks and tubs. The use of steel and enamel as a fine art medium and as a base for mural painting was pioneered in 1933 by the author.

Following wide promotion of this material, steel has been warmly accepted by artists in the USA and elsewhere.

The artist owes much to the metallurgist, chemist and technician for developing this ideal metal. Through spectroscopy and other scientific methods of analysis it was found that there are as many as twenty-five various elements in iron: boron, manganese, potassium, sulphur, titanium, nickel, arsenic, molybdenum, lead, tungsten and even gold, silver and platinum – to name a few. A few years ago boron was considered an impurity in steel; today steel-makers have learnt to control it, and boron is now a valuable and important hardening agent.

REQUIREMENTS FOR THE ARTIST

Enamelling stock sheet steel should be specified when ordering. Sometimes companies offer it under their own trade name. Manufacturers are often prepared to shear a large sheet of steel into specified sizes at a nominal charge. This gives the artist plaque sizes to fit into his kiln. This high-grade metal has excellent drawing and welding properties for the artist or sculptor desiring to make stabiles, mobiles or other three-dimensional compositions. Enamels fuse well over spot-welded areas.

For the painter, a flat surface of light-, medium- or heavy-weight sheet steel in all gauges is available. From personal experience I find 18 and 20 gauge most suitable, the 20 gauge being the lighter of the two. The surface of sheet steel is treated with lubricants at the mill to prevent the metal from rusting.

PREPARATION OF THE METAL

It is usual to cover the entire metal (front and back sides) with the enamel coating, but before this can be done the steel must be perfectly clean. All dirt and grease must be removed. The metal is immersed in a boiling alkaline soap solution for cleaning. After rinsing the metal is treated with a hot solution of 6 to 8% sulphuric acid. This pickling at 140° to 160°F for 10 minutes will remove the scale and traces of rust. The acid bath also produces an etched surface which promotes bonding of metal and enamel. The acid solution should be rinsed off the metal with clean water.

The next operation is to give the metal a 5-minute bath in a hot nickel sulphate solution and this is followed by treating it in a tank of sodium hydroxide which neutralizes the acids. When dry the metal is ready to be sprayed with enamel.

APPLYING THE ENAMEL

It has long been the practice to spray or dip a dark blue enamel (called *ground* coat) on to the steel before applying the cover enamel. This cobalt type of enamel is a bond for all subsequent coatings of white or colored enamels. Recent developments in steels and enamels have led to the production of titanium white enamels which can be applied and fused directly to the steel. This is called the 'direct-on' process.

FIRING THE COATED ENAMEL

Enamels are made to fuse between 900° and 1 600°F. The type of compositions of greatest interest to craft enamellers fuse at about 1 400°F in 3 minutes.

Sheet steel is coated on both sides so that there is no exposed metal, and after the enamel has been applied the sheet must be dried and then placed upon nickel firing pins and put into the furnace or kiln. The pins are usually made red hot before placing the sheet on them. It takes 2 to 3 minutes to fuse the enamel to the metal. When the pieces cool, further enamel applications (white or colored) can be made, and the sheet fired again.

White or colored enamel precoated sheets ready for further embellishment can be purchased in almost any size required.

POPULARITY OF SECTIONALS

Sheet steel can be cut into small squares, oblongs, triangles, hexagons, and any number of other interesting shapes, many of which can be assembled on plywood backing to make murals or works of extended size. Freedom of composition can be maintained by placing the tiles together and numbering the backs to keep them in a proper sequence after firing. Tile cement can be used to stick them to backing materials.

The ambitious enameller wishing to produce large murals on huge sheets of steel may be able to persuade commercial plants to lend their firing facilities when not in use industrially. Painting and design can be executed in the studio, and the work transported to the site for firing. Sheet steel is durable and will withstand rough handling without chipping or cracking. Painted-on colors can be transported easily. Any lumps or other texture materials can be applied to the work at the firing site.

White line composition is most effective with the use of dark and light colors. By Thelma Frazier Winter.

Pink, purple and blue colors are combined with liquid gold to produce this harlequin plate by Thelma Frazier Winter.

PLATE 46

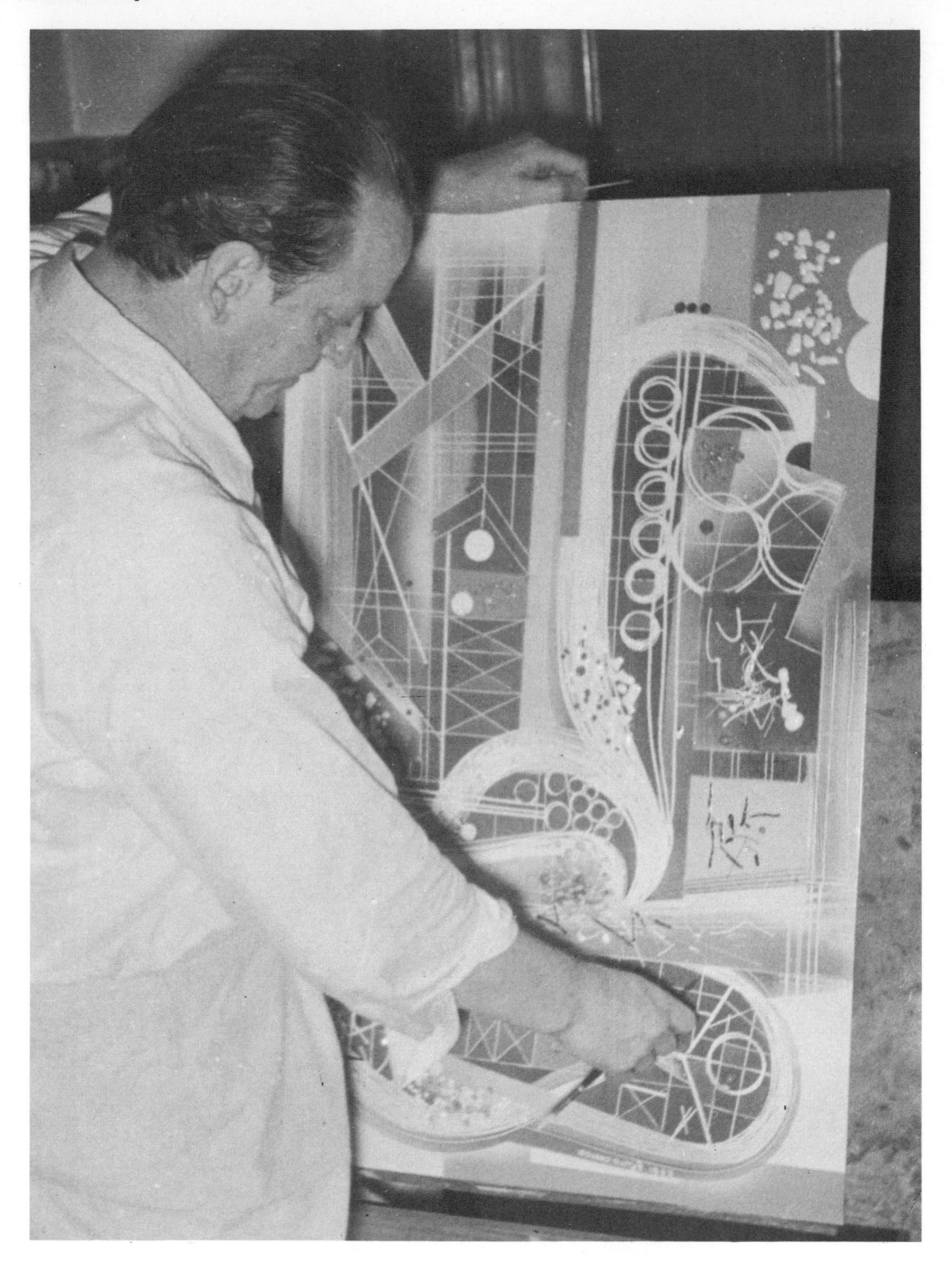

The author produces a quick line drawing through the dry enamel with a sharp pointed stick.

Working with a pqlette knife this subject shows a loose handling.

Rendering shows an impasto effect from knife application.

PLATE 48

Subtle tones in portraiture can be obtained by applying enamel with a stomp made from silk covered cotton batten.

Figures are drawn with the wide edge of a razor blade. Fine lines are drawn with a toothpick.

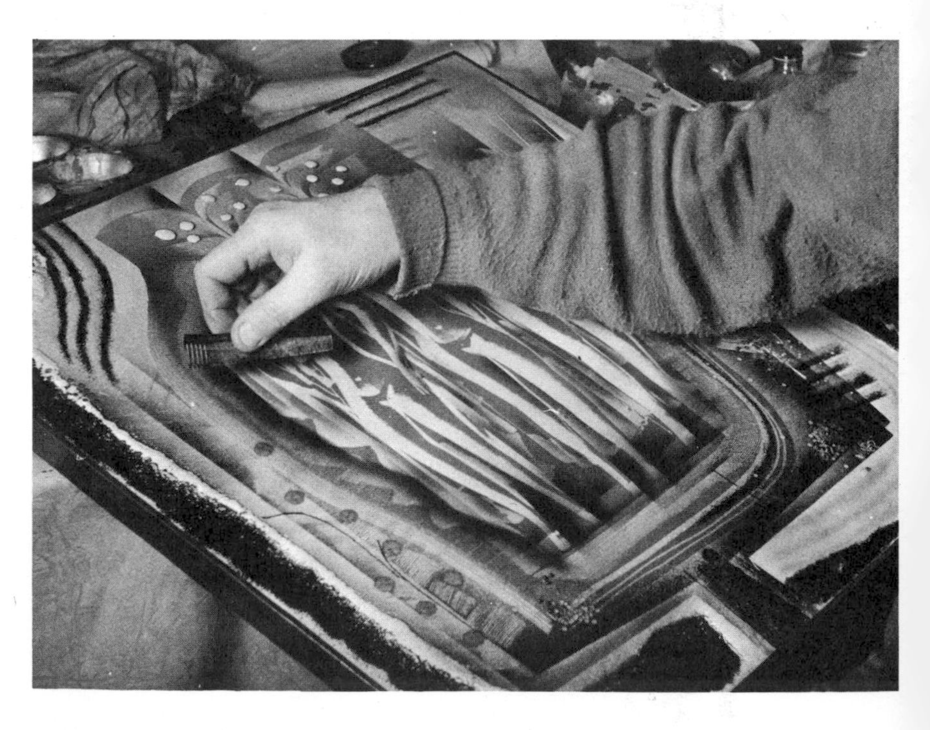

The coarse section of a pocket-comb produces rhythmic lines.

PLATE 49

Fine line graffito drawing made with sharp dental tool. By Thelma Frazier Winter.

PLATE 50

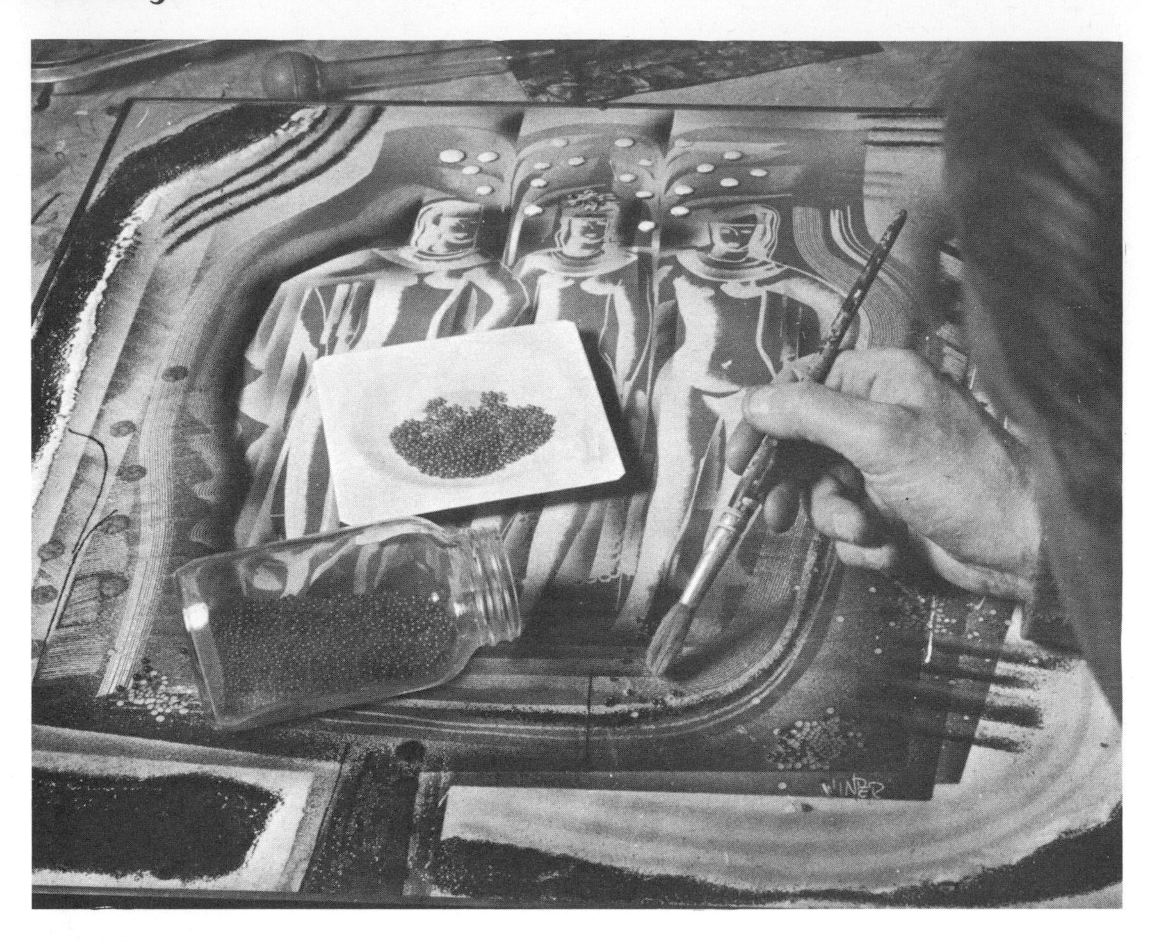

Small red-glass balls were stuck on to the surface with gum tragacanth and fired for textural effects.

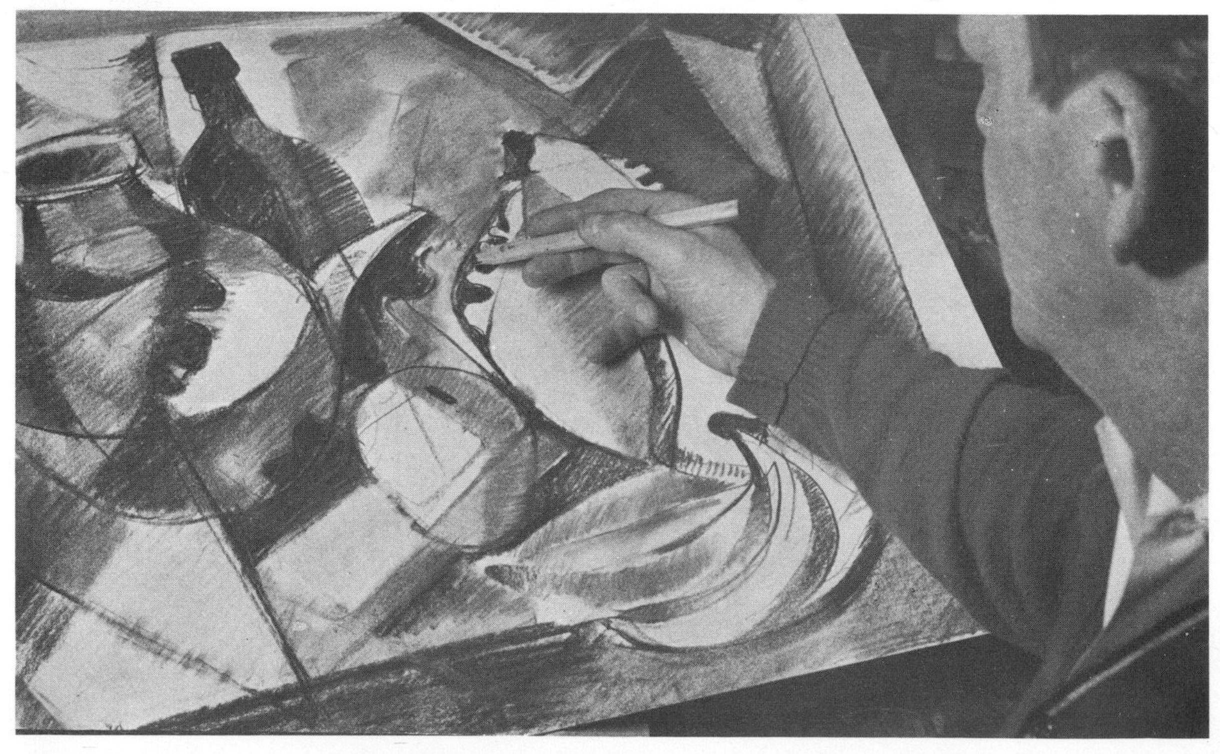

Pencil drawing for proposed still life painting.

Black enamel painted into white line drawing.

Small chartreuse and turquoise lumps used to accentuate center of flower.

The completed plate. By Thelma Frazier Winter.

Flora and fauna designs are used for these decorative plates. By Thelma Frazier Winter.

Liquid enamel poured on to panel.

Black liquid enamel applied to panel with syringe.

PLATE 54

Enamel in heavy consistency will produce raised lines.

Sheets of thin flake glass can be fused in areas to give color tone.

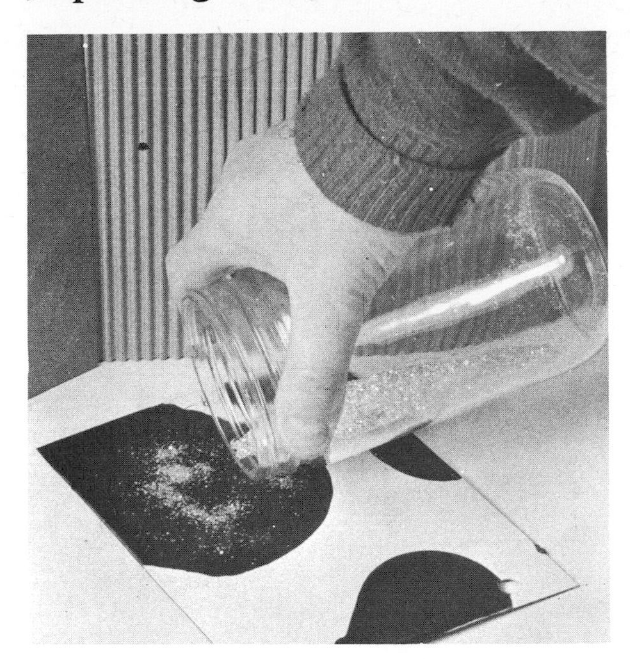

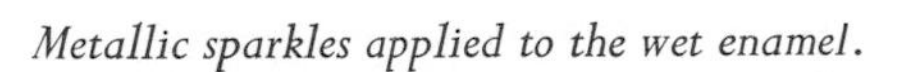

Metallic sparkles applied to the wet enamel.

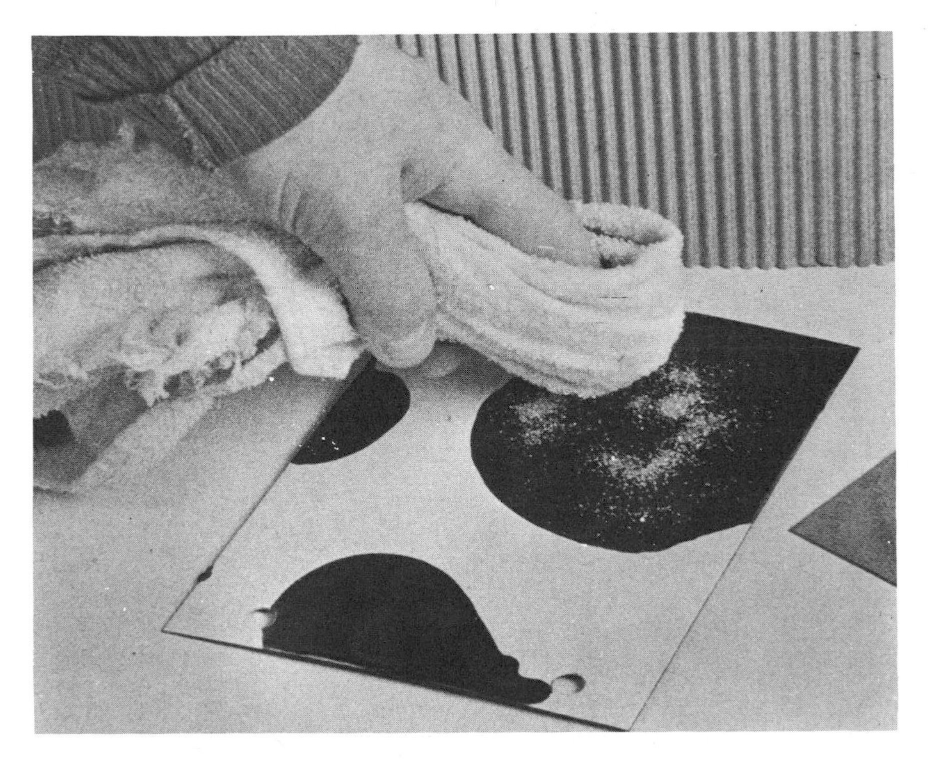

After firing the metallics can be polished.

PLATE 56

Three panels showing clear flake glass fused to bare steel surface. Dark areas show the oxidized metal.

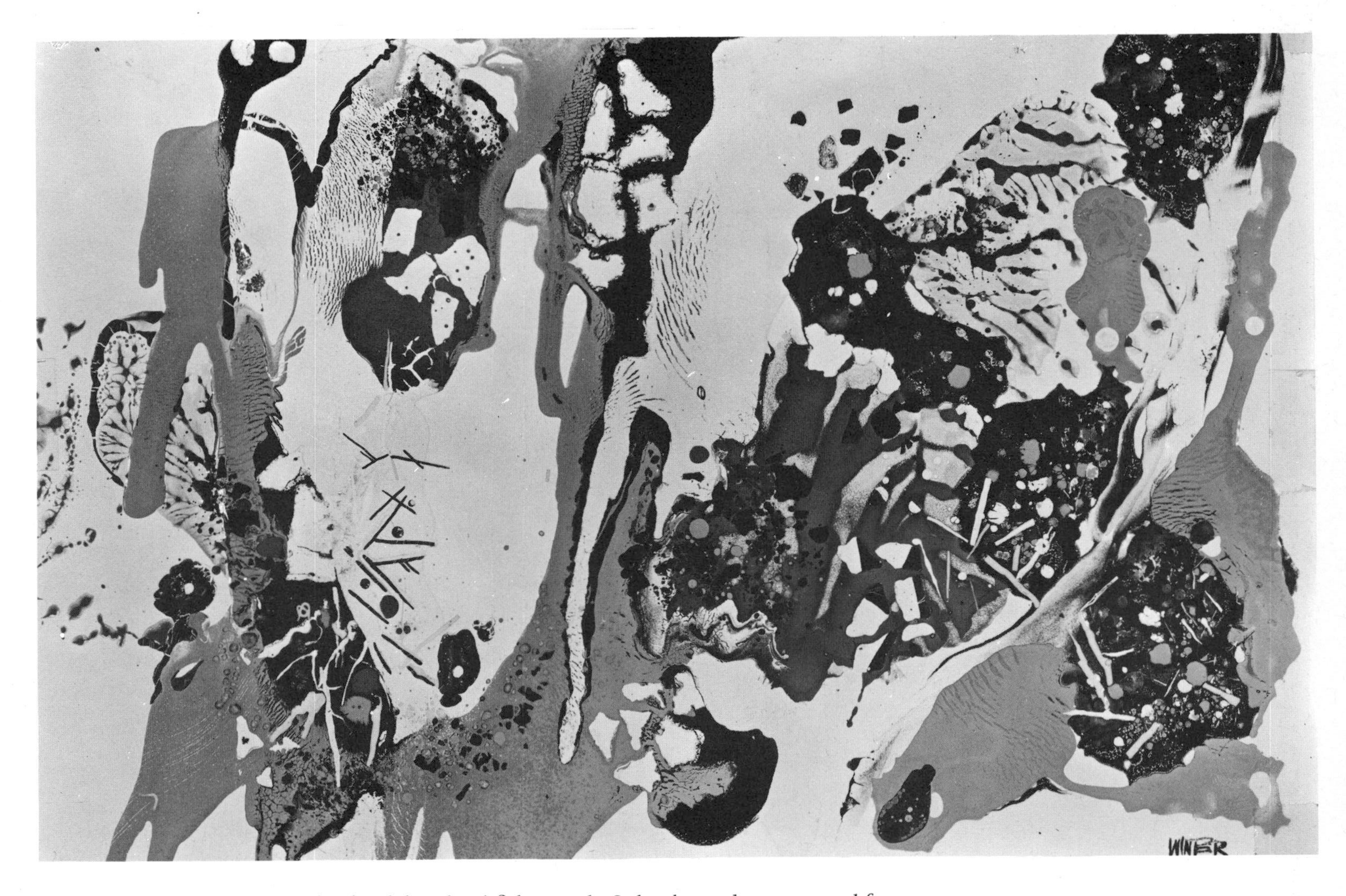

Abstract expressionist painting produced with liquid and flake enamels. Red and green lumps were used for accents.

PLATE 58

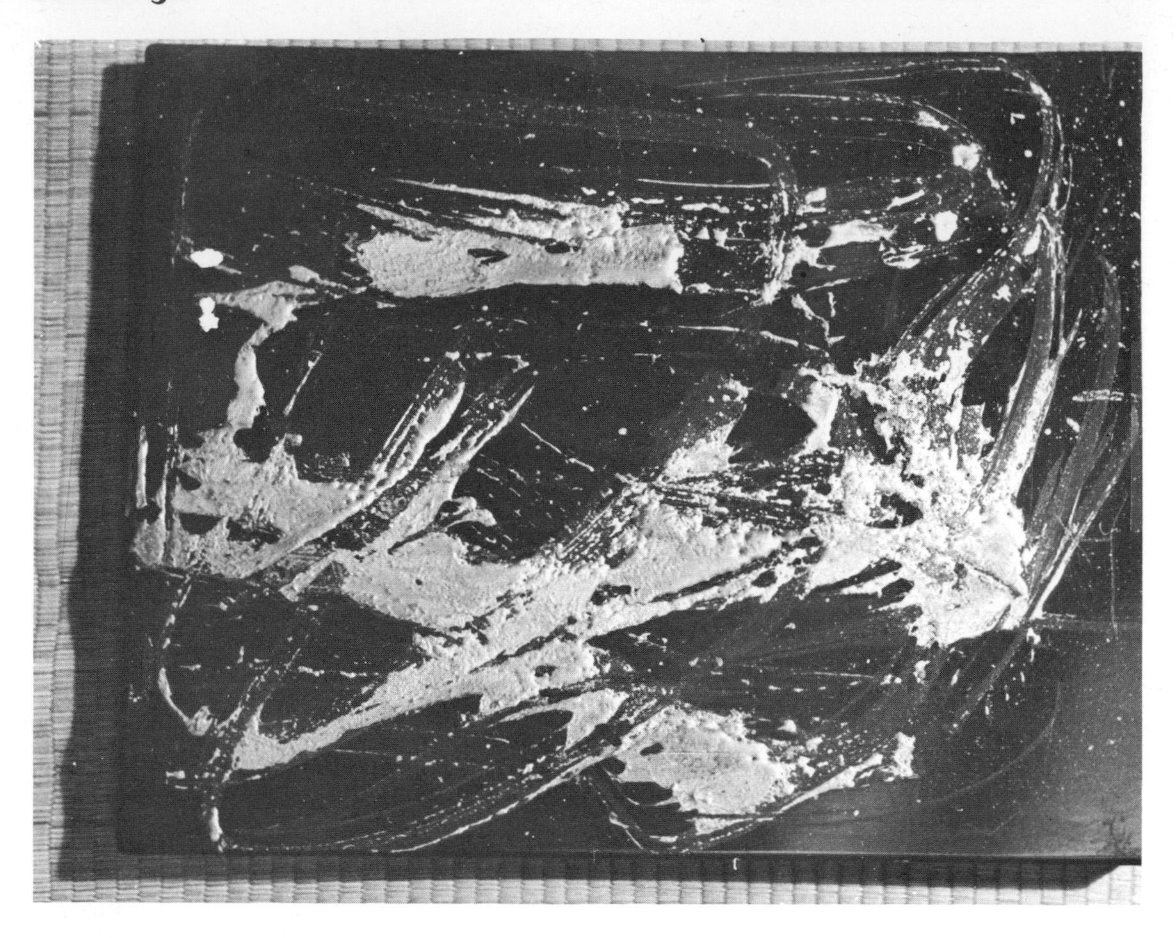

Aluminium pellets were melted on to a black enamel panel. While the panel was red hot the metal was rubbed with an asbestos glove. By John Puskas.

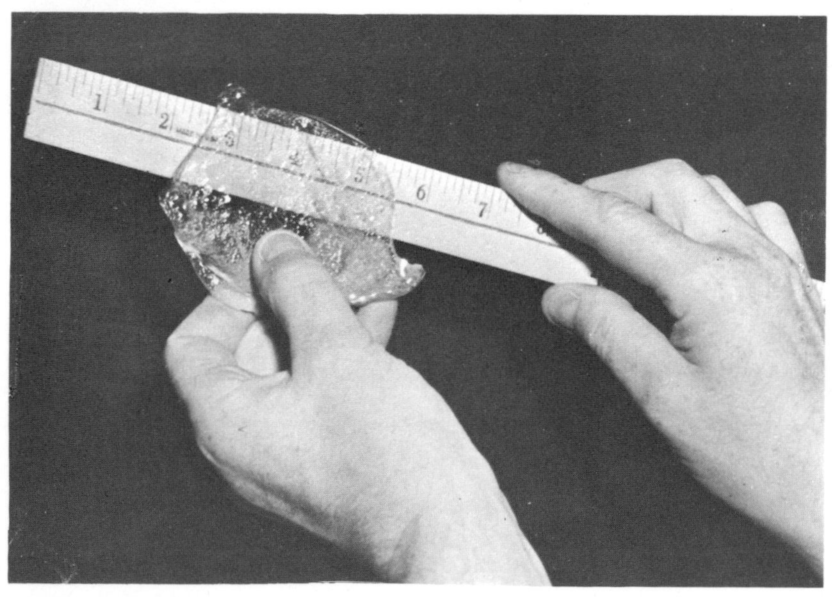

Clear flake glass has many uses.

A wide selection of roll and sheet copper and steel in smooth and embossed surfaces is available for the enamellist. Metal comes in a wide range of gauges.

Birds, butterflies, flowers and animal pendants and brooches in colorful enamel copper are the creation of Harold Tishler, New York enamellist. He has given permission to the author to show how they were made.

The horse is cut out of 20 gauge sheet copper.

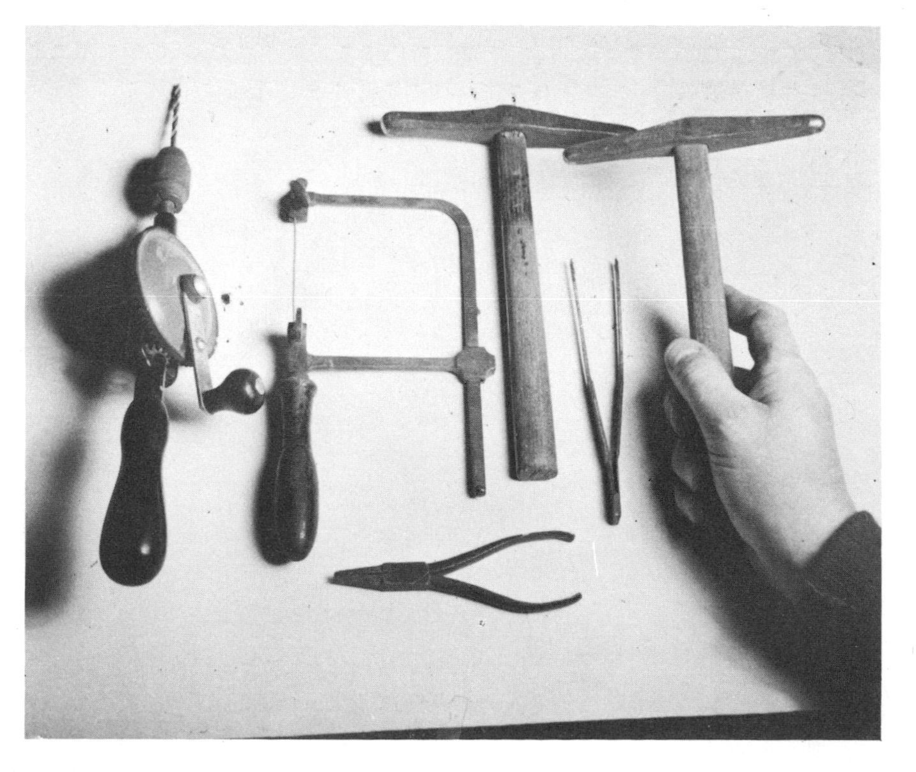

The tools for metal jewellery working are simple-shears, drills, saws, ball-hammers, pliers and tweezers.

PLATE 62

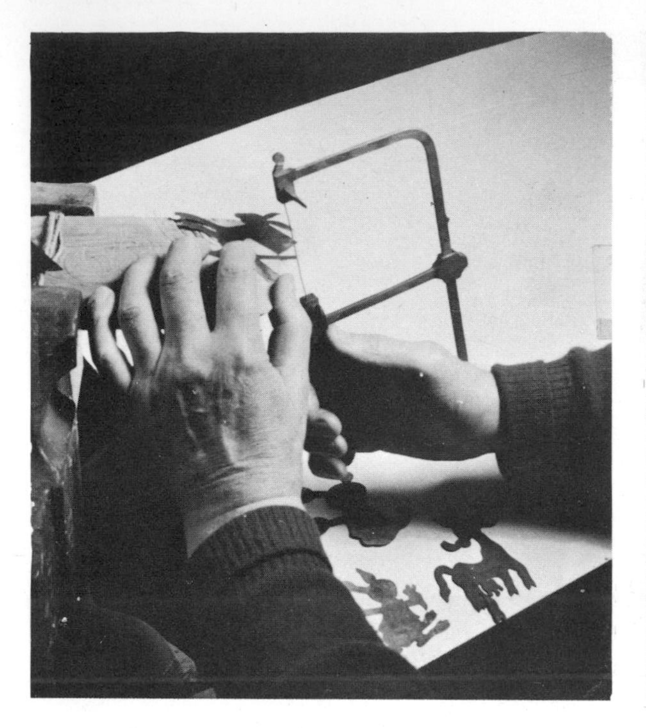

A coping saw is used to cut fine detail.

Gouged-out sections of a hardwood tree trunk are ideal for forming the metal pieces.

Shears used for straight cutting.

A good supply of color is worked-up ready for painting.

Air brush and compressor are used to apply color.

PLATE 64

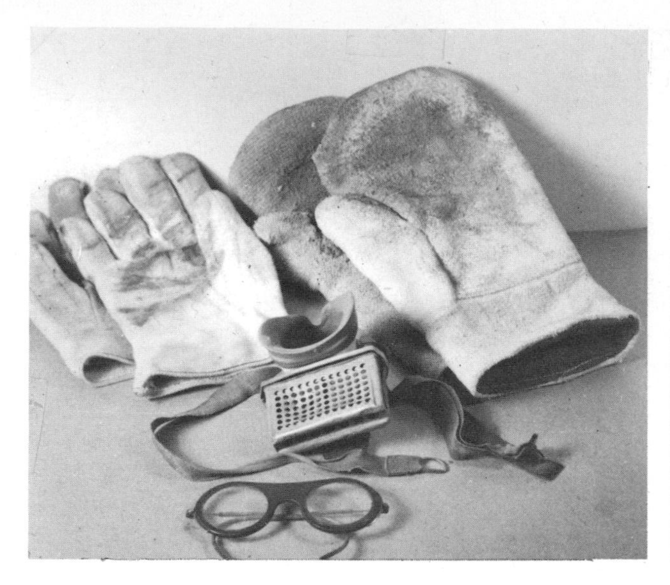

With jewellery and other small enamels, asbestos gloves and finger gloves are indispensable to the worker. Goggles and respirator are advised on larger projects.

An interval timer clock is used for firing enamels. An alarm rings at any set time.

The petals of a flower are given form.

Pliers are used to bend and twist the horse's mane.

After thorough cleaning with brush, cleanser and water the copper is dried and dipped into liquid white enamel, the excess shaken off and dried. After firing for $2\frac{1}{2}$ min. at 1 450°F it is ready to be painted with enamel colors.

Transparent color is painted on a flower.

Place enamel powder on a slab, add a few drops of oil and work into a paste with a palette knife. Add a drop or two of turpentine for thinner.

PLATE 66

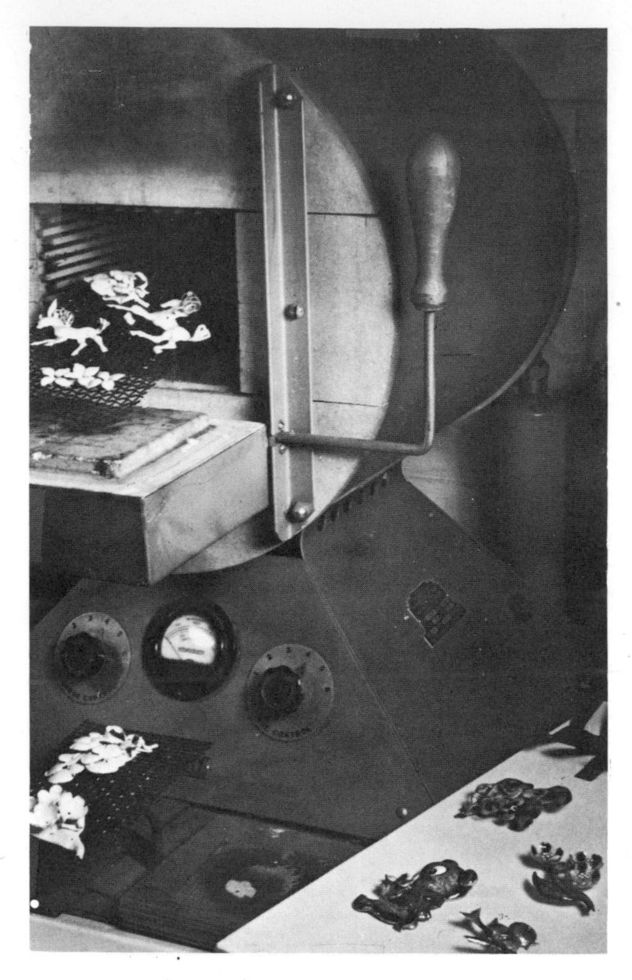

With a large chromel wire planch several pieces can be fired at once. The door can be opened at any time to examine the fusion without affecting them.

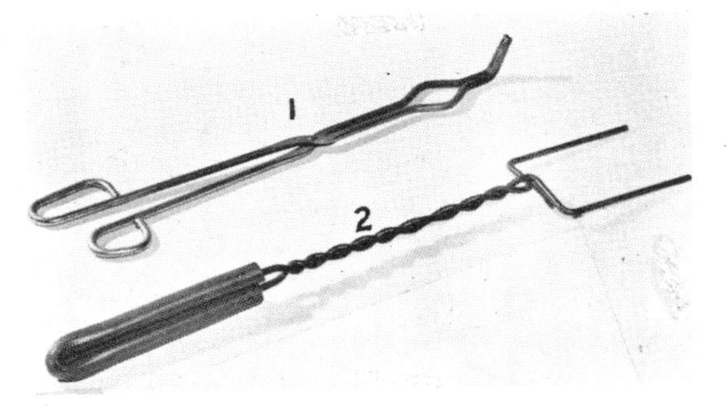

Long tongs and fork are a protection for the worker from the heat as work is brought out of the furnace.

Firing tools consist of chromel planches, supports, stilts and tongs. This type metal resists scaling in the intense heat.

The round, oval and oblong pendants are decorated with silver leaf (paillons). The floral brooches show the intricate bending of the metal and the meticulous painting of the enamel colors.

PLATE 68

Rabbit brooch, and bear-cub smelling the flower.

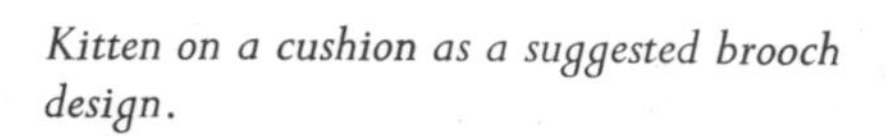

Kitten on a cushion as a suggested brooch design.

Pendants and floral pieces. Colored glass beads are used as stamens for the flower.

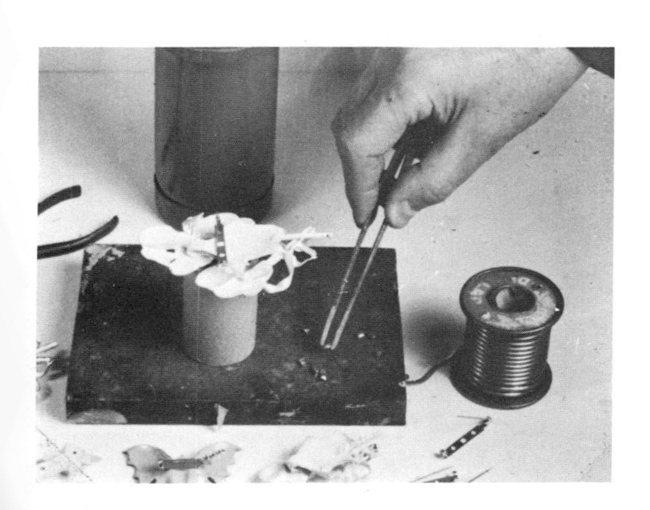

Needles can be secured to backs of brooches with epoxy cement or by leaving two bare spots in the copper and using soft solder. A play of the torch will melt the solder, but will not be intense enough to fuse the enamel.

The back sides of brooches showing the needle assembly.

PLATE 70

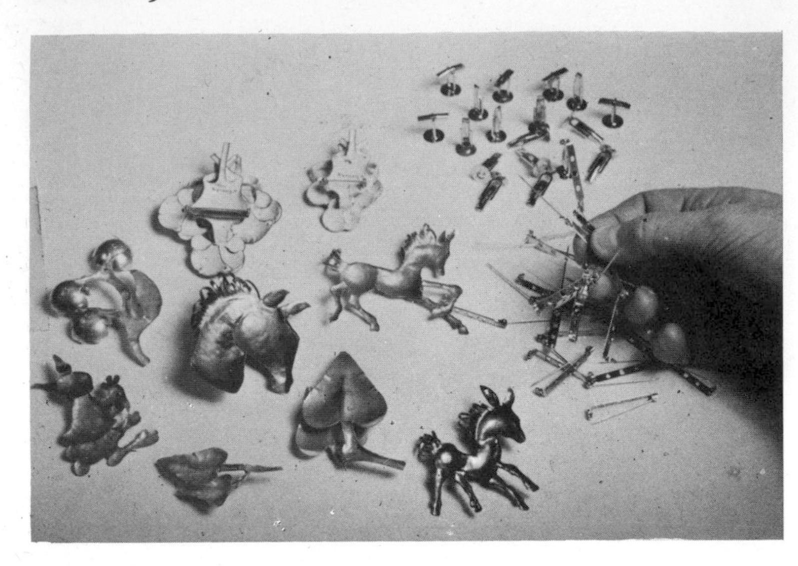

Long bar-pins were found to be most satisfactory.

Tweezers used to place bar pin.

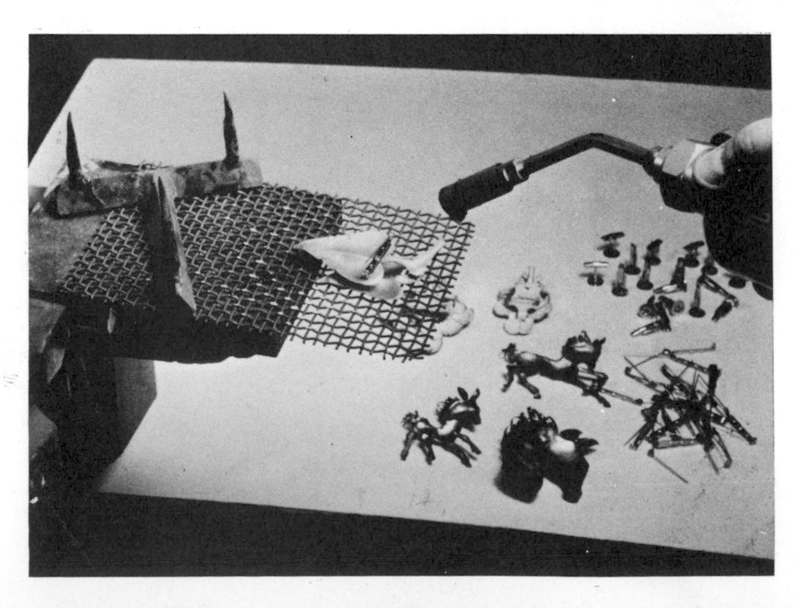

A light play of the torch will fuse the soft solder.

With smelter temperature at 2 400°F, the liquid glass enamel is drawn off with a steel rod to create textural abstraction.

PLATE 72

Black strings fall on to white enamel steel panel.

Waving rod back and forth.

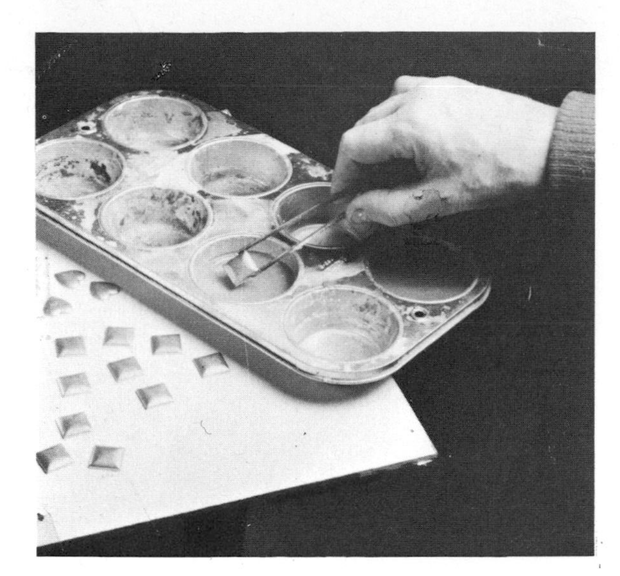

After cleaning, small domed copper squares are dipped into liquid enamel.

Pans hold an assortment of colors.

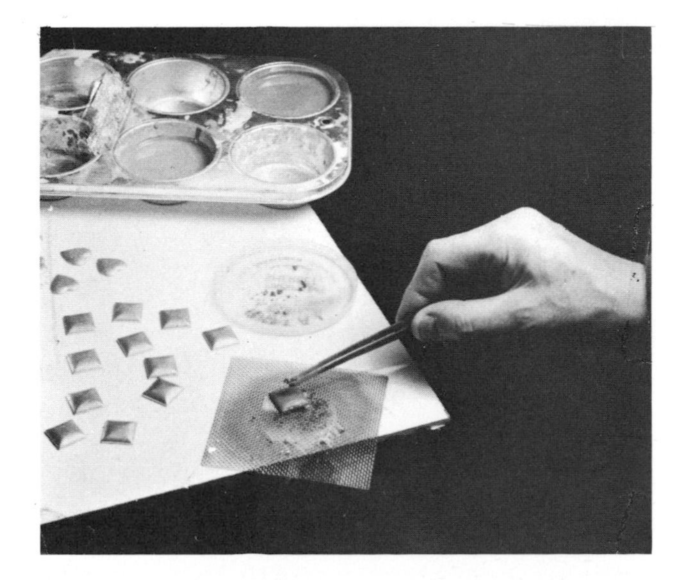

After shaking off excess enamel the square is placed on a chromel screen and dried ready for firing.

After fusing a transparent brown on the square, a turquoise lump is placed in the centre and melted upon second firing with the torch.

Small enamels can be successfully fired with a propane torch. With the flame directed at the underside of the piece the enamel will melt within two to three minutes. Large torches can fuse larger surfaces.

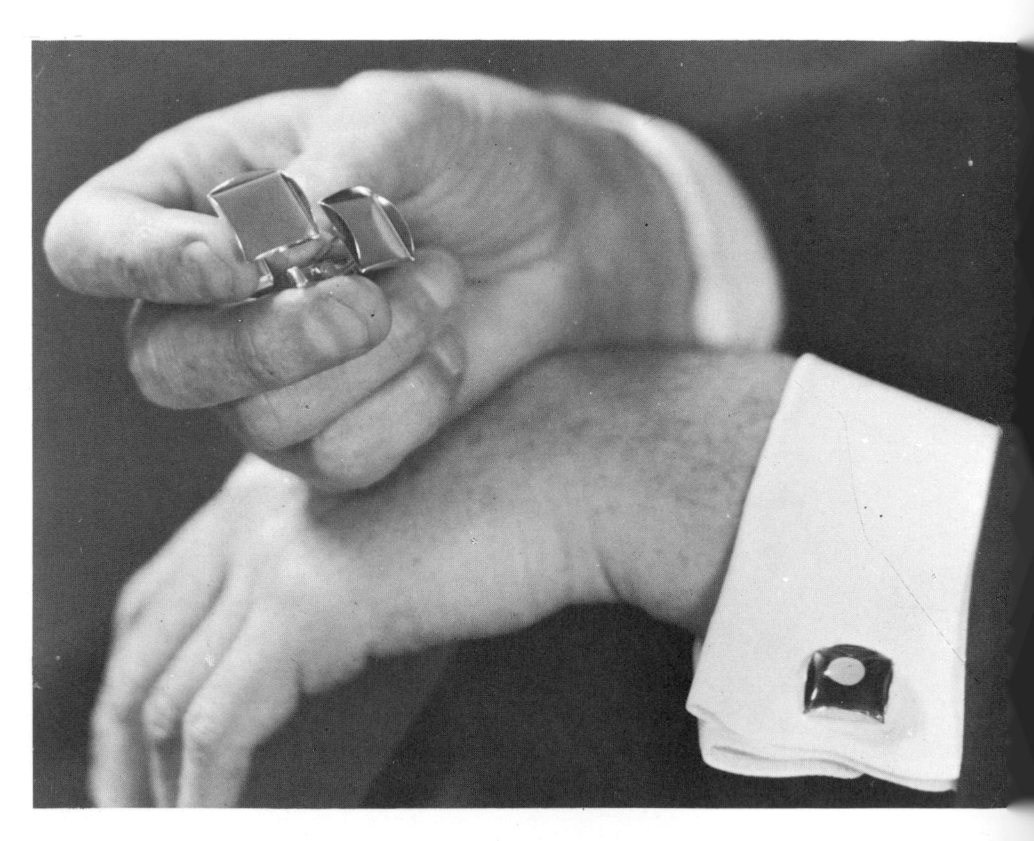

Gold plated cuff-links make solid frames for enamel squares.

PLATE 75

Silver foil can be purchased on square sheets.

Place foil between two sheets of paper for easier cutting.

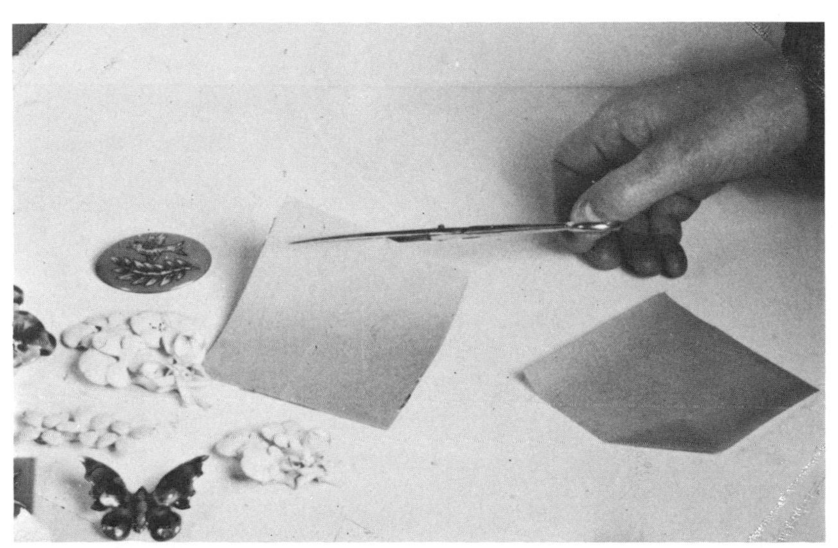

Small pieces of foil are adhered to the white enamel flower using a touch of gum tragacanth.

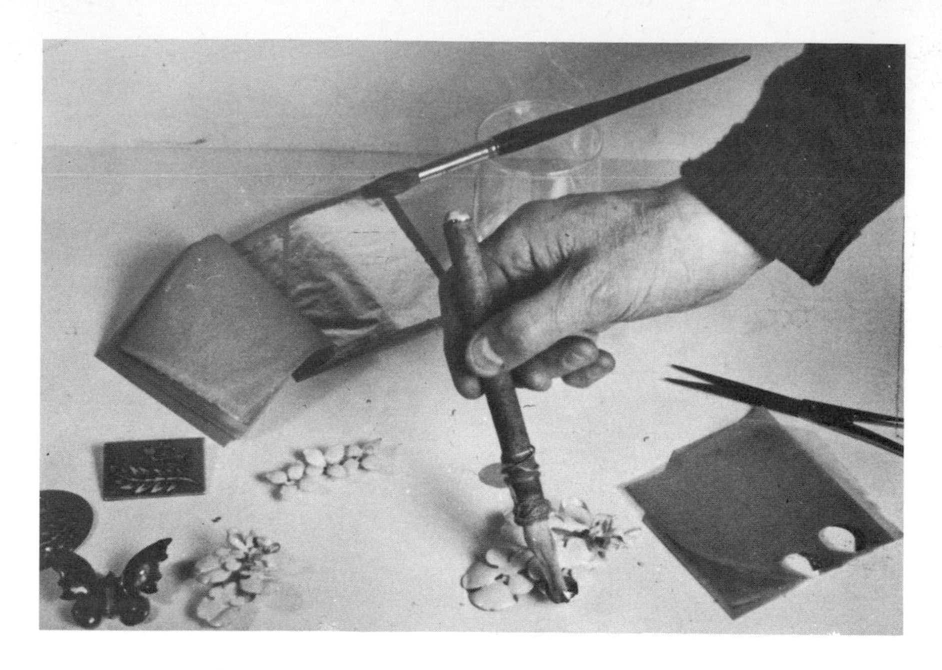

Glass brush is used to stroke the foil as it comes from the intense heat.

Tiny needle holes pounced into the foil will allow any excess air or moisture to escape in firing.

Scrap metal derived from cutting out ash trays and bowls has potential for small sculptural forms.

Metal can be shaped by use of fingers and pliers.

PLATE 78

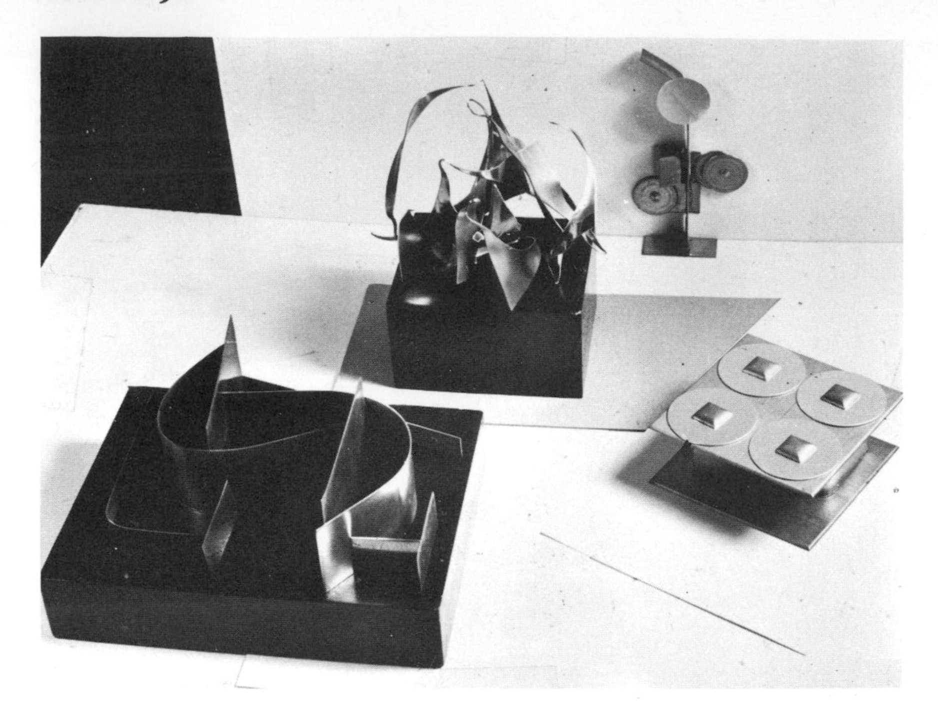

Metal shapes can be silver soldered prior to enamelling.

Wood blocks add to their third dimensional effect.

Match safe can be made by hammering copper with a flat edge hammer over a straight edged steel stake.

Enamelling process is similar to that for ash trays.

Paper stencil or template used to apply white star.

Decorative pieces with circular and radiating lines.

PLATE 80

Copper machine shavings from a lathe make interesting brooches when enamelled. When thoroughly cleaned from oil and grease they can be dipped into the opaque white enamel, excess shaken off, then dried and fired.

Bands, circles and discs of copper may be enamelled in this manner.

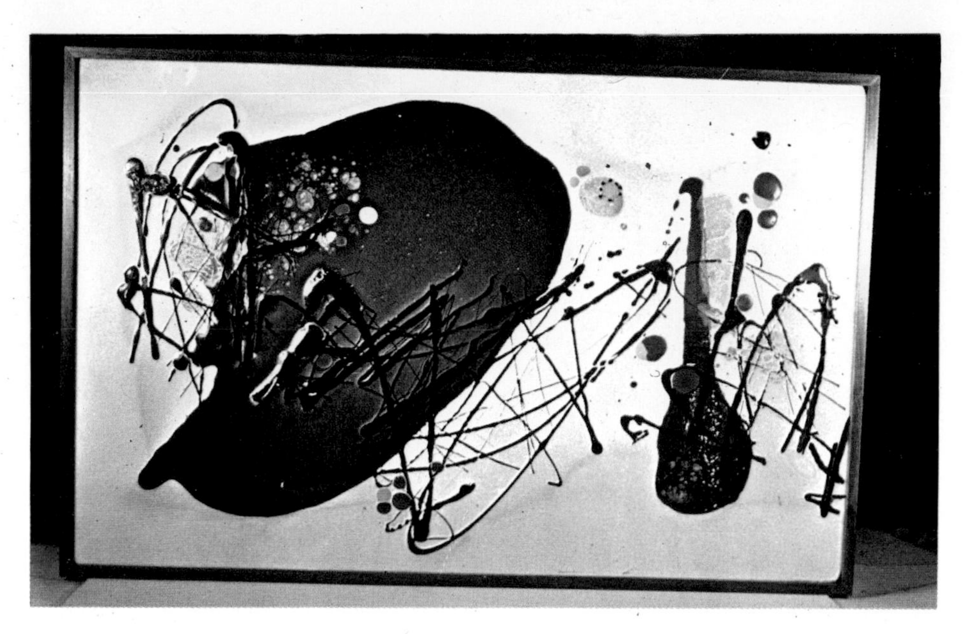

Abstract panel executed with black strings and red accents.

Fast action is necessary before the glass hardens.

PLATE 82

Sunflower enamel plates. By Thelma Frazier Winter.

Decorative accessories are made with stencils and butterfly motif.

PLATE 83

Spontaneous line drawings make interesting jewellery projects.

PLATE 84

Oblong and square shapes of metal lend themselves to pins, earrings and cuff-links.

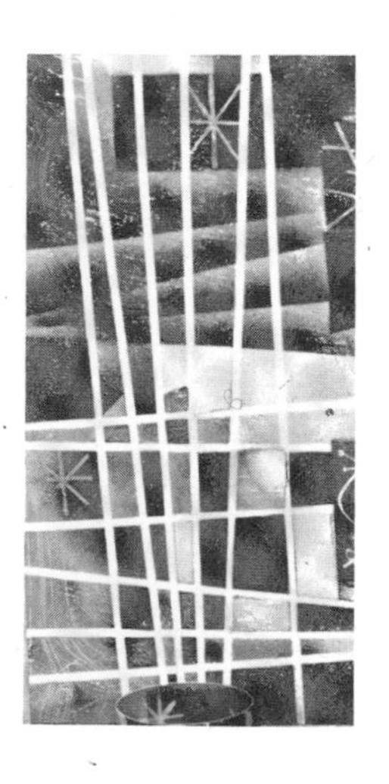

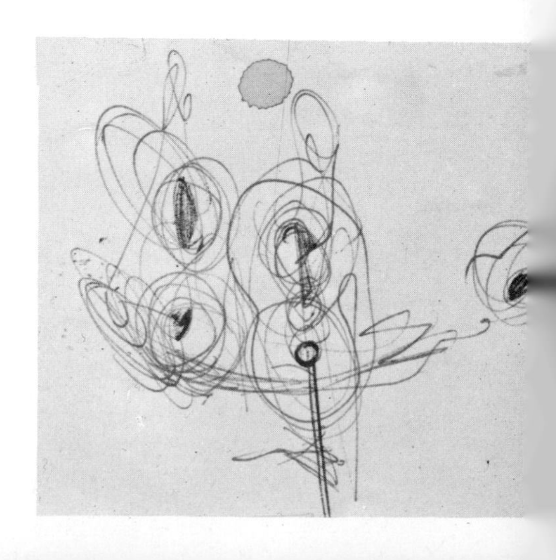

Quick action with a pen will produce rhythmic, imaginative ideas.

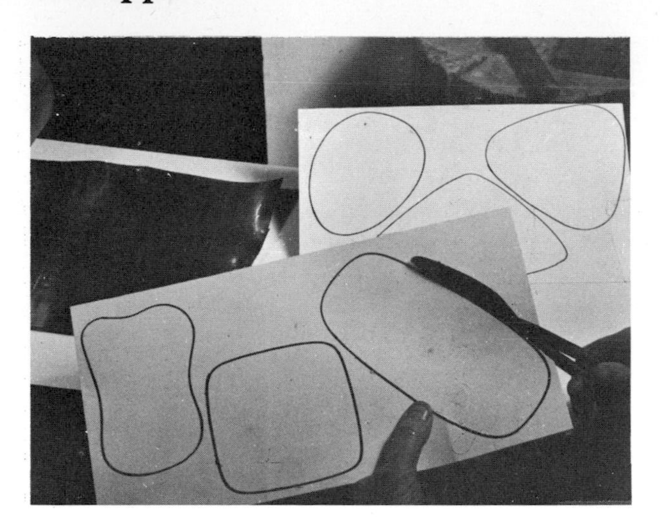

Bowl and ash tray shapes drawn on thin cardboard and cut out with shears.

Shapes traced on to 18 gauge sheet copper.

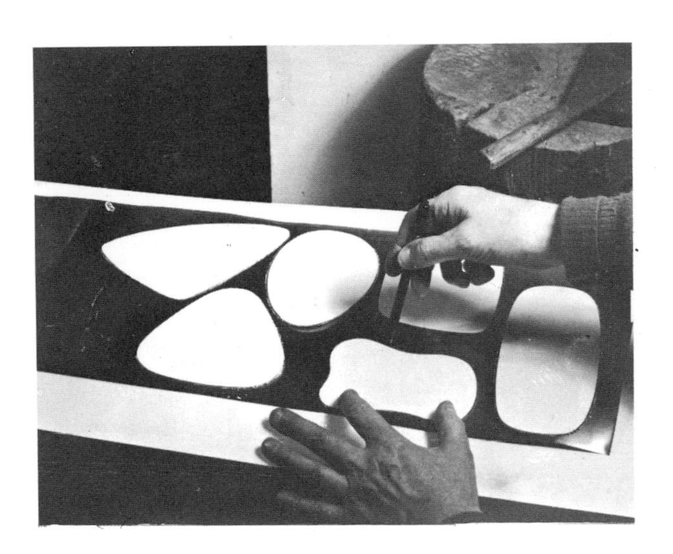

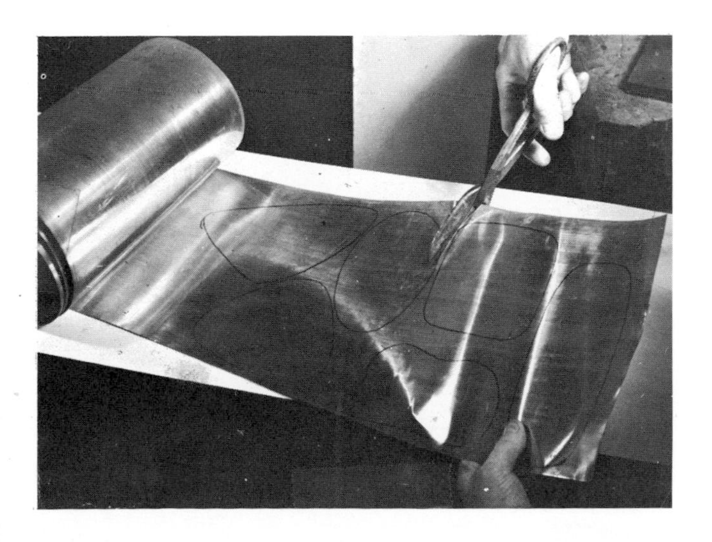

Metal shears used to cut the metal.

PLATE 86

Metal is raised by pounding it into the gouged-out section of wooden log.

The plannishing hammer shapes metal over a steel stake.

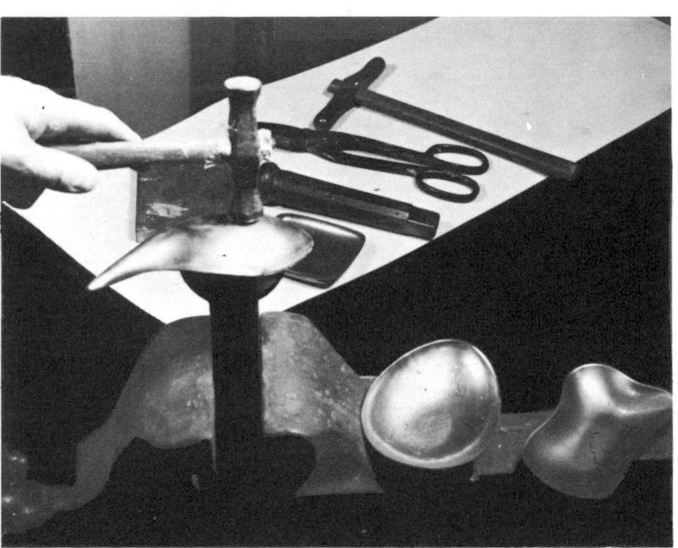

Truing metal edges.

etal shapes must be free from dirt, grease or finger marks.

After metal shapes have been properly cleaned they can be coated by dipping into opaque white enamel, shaking off the excess and drying.

Ash trays and lighters in white and gold decoration.

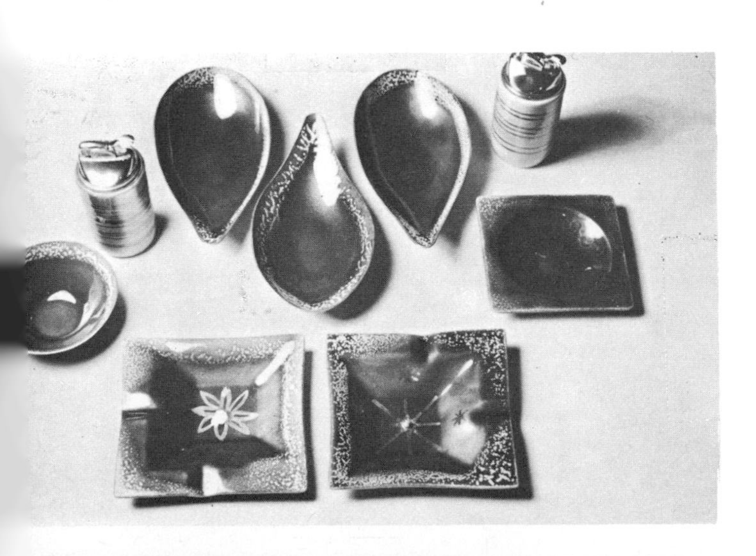

Ash trays and lighters with transparent colors and speckled edges.

PLATE 88

Footed bowls, plates and plaques with transparent speckled edges and line designs.

Metal working shears, drill, file and assorted hammers and mallets are used for metal forming.

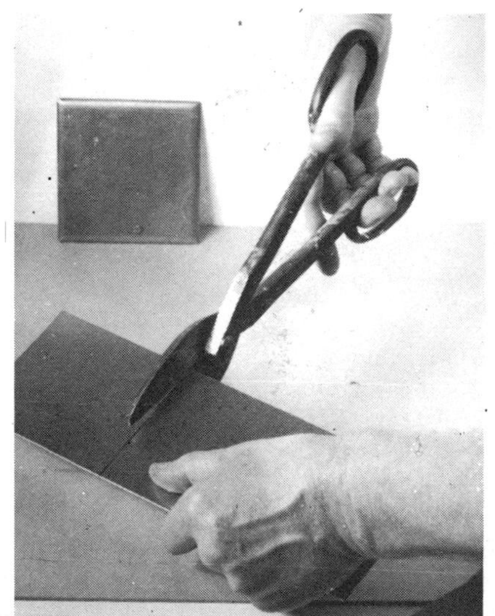

Heavy shears are used to cut 22 gauge sheet steel into squares.

Raw-hide mallet is used to bevel the edge of a tile.

Steel rod is placed in vice.

PLATE 90

Truing up edge of the tile.

Fine steel wool, scouring powder and water are used to clean steel.

Tile is dipped into a blue ground coat enamel.

After drying and firing for 3 min at 1 450° t 1 500°F white or colored enamels can be sprayed on to tile.

When color areas have partially dried, line drawings can be made with the point of a knife or sharp stick.

Free brush strokes are easy with a long pointed camel's hair brush.

Transparent enamel bowls and trays on copper by the author [1968].

Transparent enamel bowls and trays on copper by the author [1969].

hite as well as colors can be successfully pped on to ground coat surface.

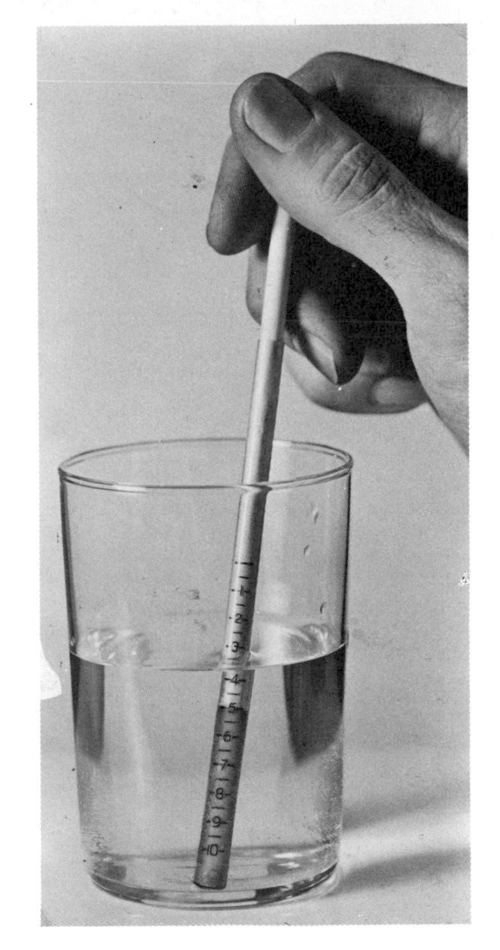

A whitish-flake gum tragacanth can be boiled in water and strained through cheesecloth to produce a clear adhesive.

hite enamel powder can be sifted on to m surface using an 80 mesh brass sieve.

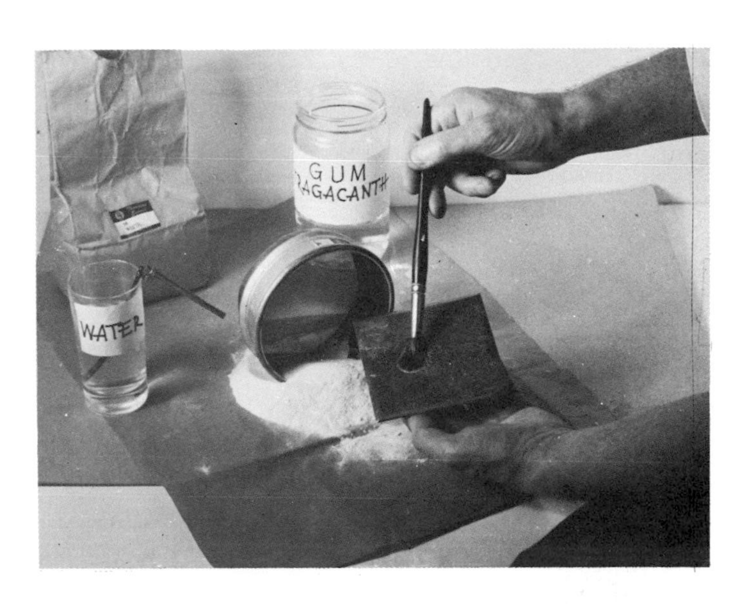

Gum is applied with large brush.

PLATE 94

Using a palette knife, the dry powdered enamels can be worked into a paste using a few drops of oil, with turpentine as thinner.

Colors are applied to a white enamel surface with brush or palette knife.

PLATE 95

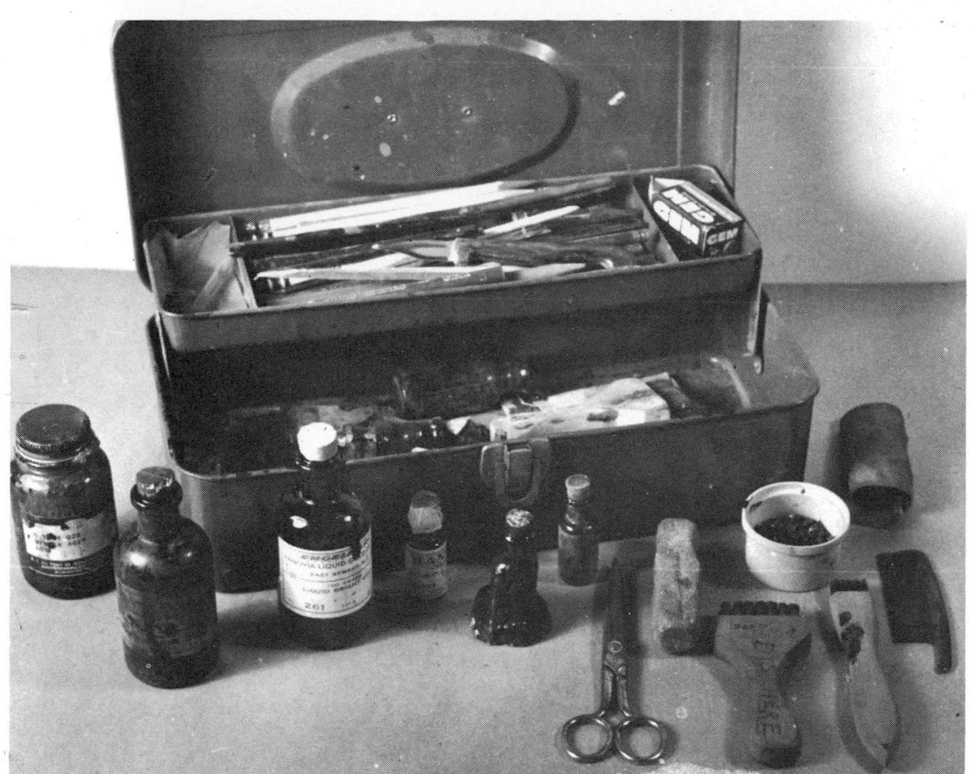

The artist's paint-box includes a wide assortment of brushes, knives, razor blades, bottles of liquid gold, silver and platinum, carborundum rubbing stones, emery cloth, scissors and pocket comb.

Brushes and tools for the execution of almost any type of project.

PLATE 96

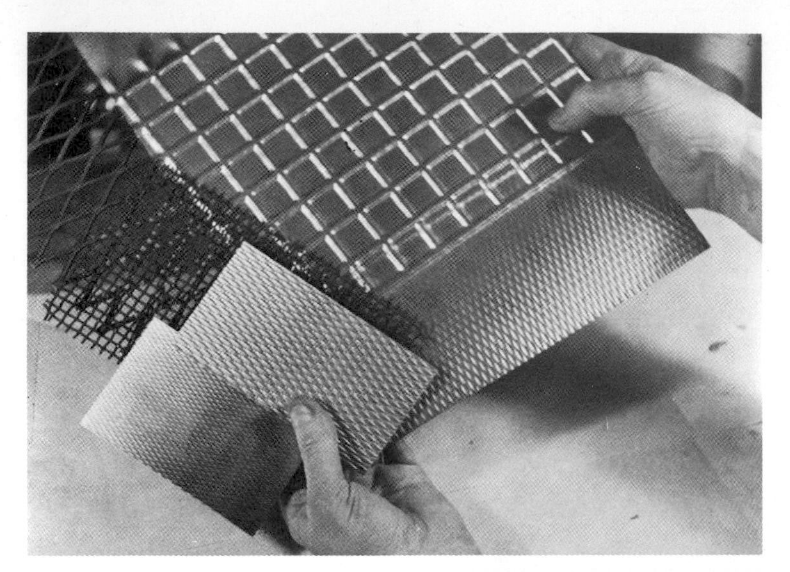

Rigidized embossed or third-dimensional steels are now available for enamelling.

Applying water to a white enamel steel panel will cause enamels to flow together.

Spoon on the opaque red enamel.

Pour black next to the red.

Other colors can be quickly applied.

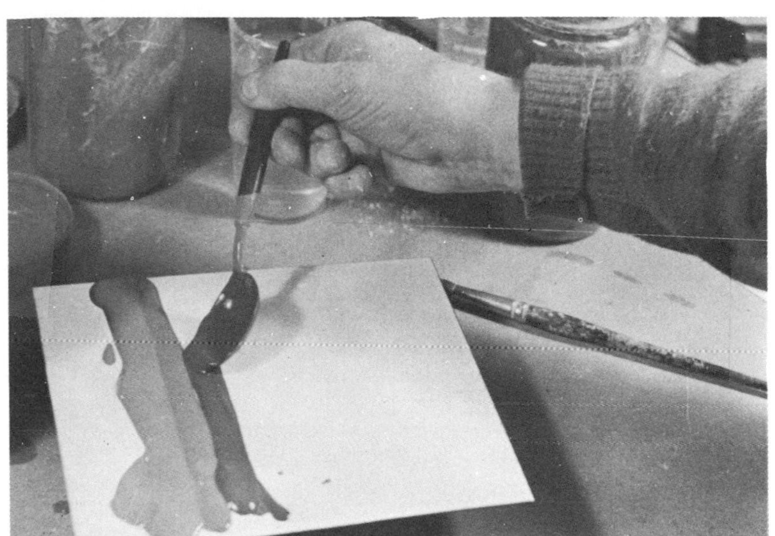

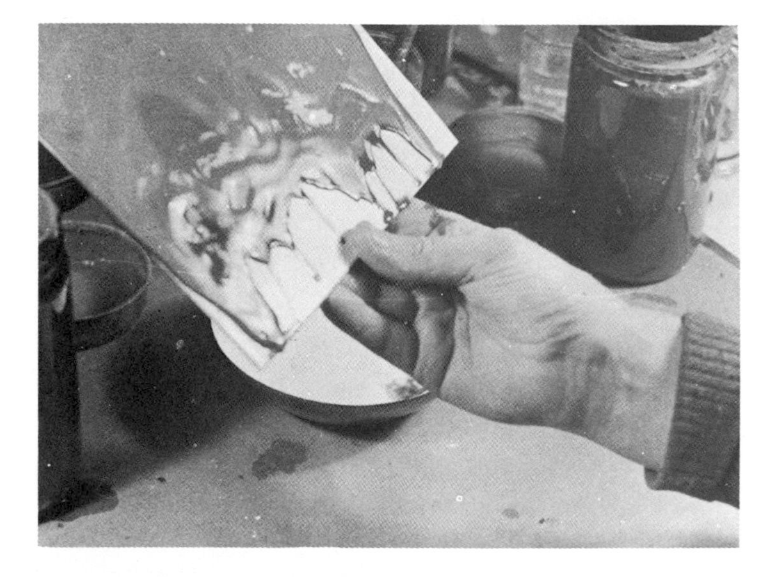

By tapping the plaque from the under side a few times and tipping it upright, the colors run together making an interesting pattern.

PLATE 98

Blowing through plastic straws produces unusual effects.

Lumps of enamel can be added for textural effect.

Small clay crucibles enable the artist to melt enamels for string textures.

Black and white strings are melted on an enamel panel with small hole perforations.

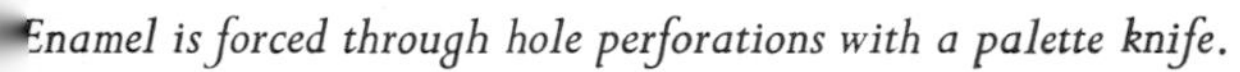

Enamel is forced through hole perforations with a palette knife.

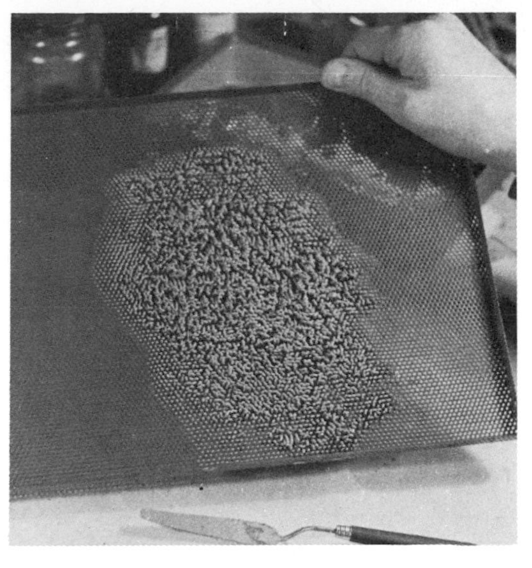

Effects of the enamel from the opposite side.

Black enamel is painted in background.

PLATE 100

Lump enamel is used for accents.

A cardboard template is used for circles.

Decorative enamel steel panels by Thelma and Edward Winter [1966].

Enamel steel panels and sectionals by Edward and Thelma Winter [1959].

Large stencil brush used for background application.

Preliminary color sketch with decoration.

White birds are outlined with liquid bright gold.

A yard stick is used as straight edge in producing straight gold line.

Smoothing the colored surface with a wide camel's hair brush.

A painted enamel steel decoration to relieve a large expansive brick fireplace. Executed in a single cut-out section of 18 gauge steel the panel is mounted 2 in from the bricks. Executed by the author.

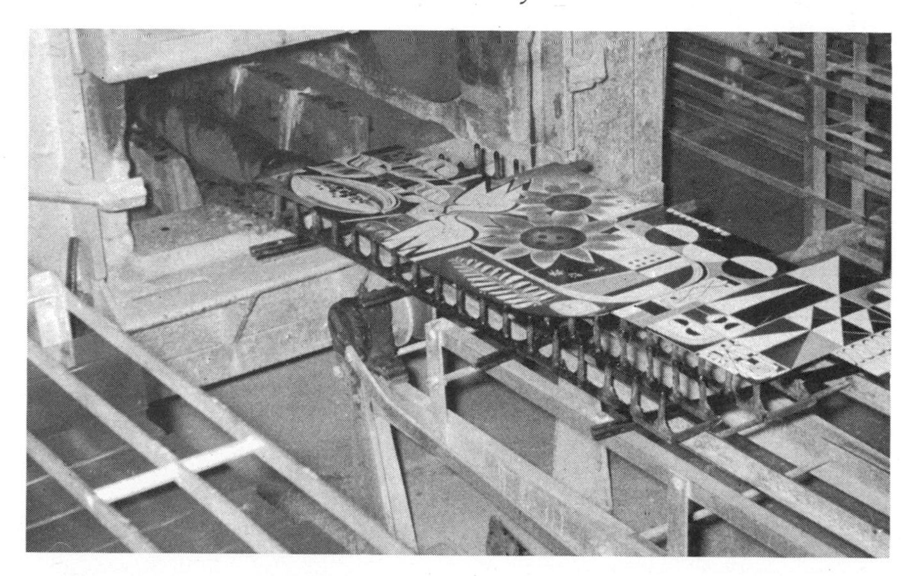

Supported on chromel firing pins the decoration is guided into the 1400°F temperature of the furnace. Firing time is 2½ min; six separate firings were required to complete the work.

Madonna—enamel and gold by Thelma Frazier Winter [1963]. *Burning Bush by the author* [1963].

Twisted band of steel with fused enamel textures.

Application of black lump enamels.

PLATE 108

Steel sheets are 'pickled' in huge tanks of caustic soda and metal cleaners.

After steel has been dried, a coat of ground coat enamel is sprayed on both sides of the metal.

Upon firing at 1500°F for 3 min, a white titanium enamel is sprayed over the ground coat and again fired at 1440°F.

Enamel is applied on large surfaces by use of the spray gun. Masking tape protects surface.

Strings and textural material must be applied to panel in flat position. Details can be applied when panel is vertical.

Floral decoration executed with use of template and pocket comb graffito – black, green, grey, pink and yellow.

13 Enamel in Architecture

GLASS enamel has had an important function in architecture for many years. Today it is recognized as an ideal surfacing material for all types of interior and exterior walls. Millions of square feet of enamelled steel curtain wall panels have been used on the exteriors of skyscrapers. It has often been wrongly promoted as a substitute for older and more familiar materials such as brick, stone or wood, but it should be recognized and used for the sake of its own peculiar characteristics which give it a unique place in the arts of building decoration.

This unique material, glass fused on metal, has all the virtues of fine china combined with the structural strength of steel, copper and aluminium. There are no set limits to the choice of color, design and surface texture. Once fused at 1 400° to 1 500°F (520 to 560°C), its surface appearance is fixed for an unlimited life span. Technical research has added acid resistance to a durable, weatherproof and fireproof material. This makes enamel impervious to all ordinary acids encountered in city air or in the average chemical laboratory. There are now ways of bonding a layer of insulating material to the inside face of the enamelled surfaces. The sun's heat, the winter's cold, and the noise of city streets cannot penetrate the space it protects.

Today many architects are turning away from the severe modern architectural forms and seem ready to employ the fantasy and imagination that artists have kept to themselves too long. Nineteenth-century industrialization was largely responsible for bringing so many new techniques and materials to the artist and the artisan.

The creative artist today abhors the earlier nineteenth-century mechanical method of detailing and copying historical ornament. He does not want to become a mechanical robot working for architects. A fine building is the most living place where art can be seen, used and appreciated. It is a far more natural place for art than a gallery or art museum.

Having been trained as an artist, it was my good fortune to produce the first enamel steel mural to be a part of a building interior [1933]. Working in the development laboratory of Cleveland's Ferro Corporation I was able to see and take advantage of scientific discoveries in metals and in a variety of glasses to fit most metals, including aluminium.

A few of the enamel decorations executed for architecture are reproduced in this book. To include them all would require a separate volume of enamels in architecture.

In recent years the Church has again become a patron of the enamel artist and during the past twenty years my wife and I have executed several huge church decorations in enamel, as well as colorful inserts for shrines. The largest enamel steel decoration in the world is the project for St Mary's Romanian Orthodox Church, reproduced in this book.

It is a great compliment to the medium to find other artists throughout the world finding the joy and satisfaction of working with enamel murals. Had the creative artist not found the industrial laboratory, this development would probably not have taken place in our time.

ST MARY'S ROMANIAN ORTHODOX CHURCH

Dedicated late in August 1960, St Mary's Romanian Orthodox Church, located at 3256 Warren Road, Cleveland, Ohio, was built at a cost of over $500,000. While modern in feeling, the church is reminiscent of the wooden churches in the Carpathian region of Romania, with steep roofs to ward off heavy snows, and the 100 ft (30·5 m) steeple to signal from in time of trouble.

The height of this church is equivalent to a nineteen-storey building – the roof 55 ft, the steeple 100 ft and the cross on the top 12 ft. An off-white glaze brick was used for the front and back of the building, and the roof is of heavy dark grey slate shingles. By using glazed brick, slate and vitreous enamel steel, the exterior of the church can be kept clean by occasional washing. The enthroned Christ decoration is 12 ft × 16 ft and composed of twenty-five flanged steel panels, each weighing approximately 21 lb (10 kg). Each panel was fired about seven times at a temperature of 1 450°F. A transparent base glass enamel was used to produce luminous light and dark blue tones in the robe, and a rich deep burgundy was used for the upper section of the robe. The drawings for the entire St Mary's project were made by Thelma Winter. Both she and the author collaborated in transposing the drawings into the glass enamel.

On the rear of the church the artists executed a huge Christ and ten figures of the Redeemed. The colors were warm red, beige, grey, black and 24-carat gold. The author's Ferro Corporation facilities enabled him to use the huge furnaces for firing the work. To date this is the largest church enamel in the world.

ENAMEL STEEL MURAL AT COLBY COLLEGE, WATERVILLE, MAINE

The enamelled steel mural installed at the entrance of the new Physical Education Building of Colby College is unique in many respects. The design combines the elements of both classic and modern styles. Sgraffito male and female figures drawn in the manner of the Greek vase illustrates the myth of Atlantis and suggests the classic origin of athletics and therefore their academic significance. In other parts of the panel Greek dancers, architectural elements including the Ionic capital and fret combine with the abstract pattern of the playing fields of the major collegiate sports. The whole design is dominated by the Great Seal of Colby College at the top left of the panel.

The mural measures approximately 40 sq ft (37·4 m²) and is made up of four separate flanged panels for purposes of firing. Fabricated of 18 gauge steel they are provided with a 2 in flange which sets the mural away from the polished Vermont Marble wall. In the process of adding color and design each panel was fired seven times. The material is impervious to the cold Vermont winters.

When the two million dollar athletic complex was dedicated, the College president, Dr E. L. Strider, had high praise for the artists and the architects Richard Hawley Cutting and Marshall Rainey.

14 How the Professional Works

MANY professional artists working in enamels find the medium lends itself to intuitive hunches, shrewd guesses, risks, leaps toward tentative conclusion, and attempts to anticipate what will happen in their work. The art process alone develops ability in hypothesizing and intuitive thinking for many artists. Since most artists do consciously consider the qualities of the medium as they create, the art instructor should encourage an awareness and sensitivity to the matter on the part of the student. The longer the student works with the medium the more this awareness will be subconscious.

Preplanning versus Spontaneous Development

THE artist must have a complete knowledge of the medium, its potentials and limitations, before the creative process can begin. With enamels it has already been shown that there are no limitations. Preplanning can be successful if either spontaneous sketches or detailed drawings are made before transposing into enamel. The manner of working becomes personal. Using properly prepared enamel surfaces, sketches and drawings can be executed in the medium itself with color and texture being added as work progresses.

Many fine projects can be ruined by not knowing when a work is completed. Overworking, overdesigning, or making raw color and texture additions in bad taste can kill an otherwise good work of art. How do you know when a piece of artwork is finished? Many criteria can be listed, some of which are as follows:

1. Intuition – one feels or senses it.
2. Technically complete, excellent craftsmanship.
3. Requirements demanded by artist fulfilled; problem solved, idea embodied.
4. Complete expression: the design says all the artist wants to say.
5. Aesthetically satisfying design; visually successful; unified clarity; parts related to the whole; suitable equilibrium achieved.
6. Dictation by the work itself; authoritative sense of rightness; conviction; unique aesthetic personality; life of its own.
7. Comparison with one's other work; each new piece must say something new or add to previous work.
8. Slackening of interest in the doing; no urge to continue.
9. No change possible without altering the whole; anything added would detract.

10. Ideas for other work emerging; it begins to change into something else.
11. Possibilities for continuing exhausted; nothing else requires doing.
12. Enamel surface is satisfying and all metal edges or exposed metal areas are smooth and highly polished.

Craftsmanship

THE aspiring craftsman must remember the time-worn statement: 'We must crawl before we walk, and learn to walk before we try to run'. These have always been the logical steps to be taken by anyone learning to produce art. Little of value can be produced without ability, the time and patience required to get the feel of a material and to learn the secret of getting the most from it. Although this book, with its step-by-step photographs of craftsmen working with materials, can light the way the mere turning of pages and absorbing printed words will not magically produce an enamel craftsman. This process must be personally entered into by the individual after he has been shown the way with the necessary stimulating materials. Who can tell what powers lie within any person waiting the right word or the right thought to bring them to the surface? Far too many people rationalize their own inertia, convincing themselves that they cannot begin to create in a given field.

Anyone who has ever accomplished anything worthwhile has had to go through a period of beginning, just as the old masters did in the past. The professional craftsman and teacher must give encouragement and help to the beginner. By encouraging the hesitant starter the teacher gives confidence, and the student learns to handle his tools, to get to love the feel of soft pliable metal, and ultimately to experience the thrill of seeing powdered or liquid glass melt and flow over a metal form and cool into a jewel-like luminous surface.

Craftsmanship consists in designing the basic form and then translating the idea of the form directly into the materials. Each is essential to and inseparable from the other, and each sustains and reinforces the other. The learn-by-doing concept is fine, but without the designing process the true craftsman cannot materialize. Craftsmanship implies a legitimate union of both conception and construction. Most artists or craftsmen start out by trying to copy the style of a favourite master, a skilled teacher or some professional. This shows keen perception on the part of the young student – that he is eager to get on the right track. He soon discovers however that his copying efforts are not wholly honest, and he soon learns the only way to succeed as a creative individual is to be an individual, dreaming up new thoughts and ideas that are totally his own, and trying out new techniques and processes in the search for self expression.

My own attraction to enamelling did not arise from the standpoint of metal working, the knowledge of tools, and the ability to use them, but from the jewel-like appeal of the enamel itself. Since enamelling consists of fusing powdered substances on to metals it meant that my future success in the craft depended on my learning to design and work the metal itself. In learning to form, hammer, and create three-dimensional shapes in copper I soon developed a new love – the love of metal working. I now knew the pride of personal accomplishment, a deep satisfaction in transforming hard work into the joy of achievement.

15 History of The Art

Art Enamelling in the USA

IN view of the mounting popularity of enamels in the USA and throughout the world it is important to put on record an accurate account of the early growth and development of the craft.

In the USA, the manufacture of enamel frits started in the early 1880s, when two stepbrothers, J. H. Collinwood and T. D. Mayfield, came from England and established small factories in Providence, R.I., and Newark, N.J., respectively.

A. I. Carpenter, whose company later became Carpenter and Wood, established enamelling in Providence about the same time. The enamels that these companies turned out were used for jewellery, watch, clock and meter dials on copper, silver and gold. About 1900 Louis Manz came here from France, skilled in traditional techniques and commenced producing a pictorial type of Limoges enamel. Even earlier, in the 1890s Ole B. Owren, emigrating from Norway, became a *plique-à-jour* expert, producing delicate transparent jewellery without a solid metal backing.

In the 1890s, G. Hein, A. W. Beerbohn, Edward Dietz and Louis Perochet in the New York district, were producing for the commercial jewellery trade.

In 1880 William Marlow and his son-in-law, Thomas C. Thompson started manufacturing jeweller's enamel. Upon the death of Thompson, his son carried on the business in Wilmette, Illinois, until the small factory burned to the ground in February 1941. The factory is located today in Highland Park, Illinois.

In 1902 one of Thompson's customers was Charles Thomae whose production consisted of enamelling machine-turned silver parts for mirrors, brushes and other similar articles. By 1908 he had perfected and popularized engraved or machine-tooled designs on metal. Transparent enamels, fused over these surfaces, produced a luminous reflective quality, difficult to duplicate in any other material.

About 1916, L. H. Martin arrived from France and established a small shop in Lowell, Massachusetts, and accepted a number of students from the Boston vicinity. He taught the traditional 'pointer and spreader' techniques.

Although we have no recorded proof that the work of these exponents ever became part of a museum collection the contribution of these early enamel workers should be recorded.

To return to the more commercial production of enamels, Providence was an extremely active city in the jewellery manufacturing business. With jewellery production came small stampings, such as rings, buttons and lapel insignia. Then came nameplates used on stoves, machinery and bicycles. The first automobiles produced in 1900 all carried transparent enamel nameplates. Larger die stampings introduced compacts, dresser sets and powder boxes; most of these metal parts imitated machine-turned backgrounds similar to the *basse-taille* technique. It is interesting to note that the type metal used for the automobile nameplates was guilder's metal (95% copper : 5% zinc). They were engraved by the drop hammer die-stamping method. Although none of this work can be considered as art it did serve to introduce jewel type enamel to the American public.

Between 1918 and 1930 metal craftsmen practised enamelling in the commercial jobbing shops or were associated with the exclusive Tiffany type jewellery houses. The more costly objects were embellished with diamonds, rubies and other precious stones, and though these items were costly the transparent enamels were considered precious and always used quite sparingly.

As early as 1933 the records show a list of new customers for transparent enamels including Harold Tishler of New York and the present author, Edward Winter of Cleveland. These two men had returned from studying enamelling at the Kunstgewerbeschüle, Vienna, under Josef Hoffmann.

Enamel production flourished owing to the new modern application process involving the use of a brass sieve. Through this process larger areas of the metal could be covered rapidly, thus greatly extending the application of the medium.

The invitation which the author received about this time to use the development laboratories of the Ferro Corporation and their huge furnaces resulted in his pioneering large bowls, plaques and murals.

Ancient History of Art Enamels

VITREOUS enamelling of metals is one of the most ancient forms of artistic expression. Every civilization has left some of its spirit and color embodied in this jewel-like medium. One of the more amazing features of the ancients was their ability to create this transparent substance out of murky opaque minerals and quartz materials. The ability to make a heat intense enough to melt refractory minerals was feat enough, but the works of lasting beauty created by the ancients are truly amazing.

Enamels were originally used as substitutes for emeralds, diamonds and rubies. However, the craftsmen who continued to employ them were no mere imitators; they developed enamel into an art so rich that it became the most important means of decorating the various objects used in the ritual of the Church. It becomes evident to anyone who has seen the collections of enamel masterpieces in the museums of Europe and America that this art is one of the most valued heritages from the Middle Ages.

The Egyptians and Phoenicians are credited with the discovery of enamels, and one of their greatest achievements was the application of enamels to pottery and brick. The application of enamels to gold and bronze in the form of jewellery can be attributed to the Greek and Roman civilizations during the fourth and fifth centuries. The art of ancient Rome lacked the inspiration of Greece, and most of the Italian work was confined to copying Greek shapes and styles.

The Celts and Saxons became actively engaged in enamelling between the sixth and ninth centuries, and many fine gold and bronze ornaments and shields were produced by them.

From a close study of the styles, techniques, and colors used it would appear that from Ireland the art was transferred to Byzantium where it flourished for several centuries. Fine examples of the *cloisonné* technique in particular, can be found in the South Kensington Museum collection. Primary color schemes were used almost entirely, and throughout this period of Byzantine enamelling there is an almost total absence of subtle tonal colors. The subject matter, figures and other motives were treated entirely as severe decoration, without the least semblance of expression by subtle detail. The sheer simplicity of decorative design, the strength and general character, and the richness of color makes this period one of the finest the art of enamelling has seen.

The next great application of enamelling was at Cologne, Germany. Nicholas of Verdun produced an altar front masterpiece at Klosterneuburg which consists of fifty plates in *champlevé* enamel. This is just one of the many magnificent shrines in this Province.

From here the secrets of the craft were taken to Limoges, France, but history records no new methods of handling the medium in Byzantium, Cologne or Limoges.

The newest and greatest development in the art can be attributed to Italy around the fourteenth century. The process called *basse-taille*, consisted of low cut relief wherein transparent enamels were used. This technique demanded the highest skill of the artist, sculptor and enameller. Gold and silver were used as base metals. Fine line engraving of subject matter was attained, after which very transparent enamels were sifted over and fired. This technique produced subtle detail.

Plique-à-jour was discovered about the same time. This technique resembles a stained glass window. The transparent enamel is fused to a metal base and *cloisonné* wires are used to give a solid appearance to the piece. Then in the final process the thin melting backing is etched away, giving a stained-glass effect. Owing to the extreme fragility of the work obtained with this technique, very few examples are to be found today.

Enamels reached China long after they were being made throughout Europe. They were probably introduced into China by Arab traders or by travelling craftsmen working their way eastwards. The Mongolian conquest also introduced enamelling to the Chinese in the thirteenth century. Chinese enamels fell into the three popular categories: *cloisonné*, *champlevé*, and painted.

Japanese enamels cannot be dated any earlier than the end of the sixteenth century. The Western influence which promoted the art in China does not appear to have penetrated Japan. The Japanese made use of the *cloisonné* and *champlevé* techniques but the modern Japanese have simplified and modified the *cloisonné* process with remarkable ingenuity.

The craftsmanship which reached such heights in the rich floral decorations of porcelains was turned also to the making of metal vessels, covered with patterns very reminiscent of the designs on porcelain. The art of painting in enamel was also practised in China, particularly in Canton which has given its name to the modern variety produced for the export market.

With the development of painted enamels in the fifteenth century the artist was no longer confined in his subject matter or to designs that were suited only to the limited materials and techniques his predecessors were forced to use. The sharply defined areas, each filled with a single color (*cloisonné*) and the grooves and channels (*champlevé*) and backgrounds or figures engraved in metal; the low silver reliefs covered with translucent enamels – all these were the limitations of each of these crafts and the products showed the inevitable result.

The artist working with painter's enamel and pointed brush using powdered glass as his pigment and heat for his fusion, was more a painter than an enameller. Bare metal had no part in his designs. The metal used was a working base metal necessary as a support for the opaque cover coat and the colorful design overlay. The metal served the same purpose as a painter's wooden panel. By giving the metal a subservient function he destroyed the relation of the enamel to the metal, and the jewel-like effects which for hundreds of years had given enamel their special qualities.

This development led the art of enamelling closer than ever to that of painting. Panels became bigger and the compositions more elaborate. The same tendencies which characterized the schools of painting from the fifteenth century onwards may be observed in enamels. Not all craftsmen followed in the ways of the enamel painters. There were those jewellers who continued to allow the precious metals to play a part in the design. Many times *cloisonné* and *champlevé* techniques were used in conjunction with the painter's enamel to enhance the beauty of their compositions. In some of the finest works of art produced by the goldsmiths of the Renaissance, the discrimination with which enamel is used shows a high appreciation of its qualities.

One of the earliest artists to work in the painter's enamel technique was the Frenchman Nardon Penicaud. One of his early pieces dated 1503 can be found in the Cluny Museum. His popular work won him many followers, including Leonard Limousin of Limoges [1532–74] who specialized in portrait painting. Examples of his work of this period can be found in many European and American museums as well as in private collections.

Pierre Reymond also produced many fine works in this painter's *grisaille* technique, and during the sixteenth century Henri Toutin and Jean Petitot, miniature painters, became very popular.

In England, the painter's enamel art was practised in Battersea and South Staffordshire in the eighteenth century. The enamelling produced at Battersea consisted of landscapes, figures, flowers, birds and portraits of celebrities. These were either copied from paintings or adapted from prints of that period. Transfer printing was tried successfully, the required designs being engraved on heavy copper plates that were capable of transferring about 200 successful impressions to paper in adhesive ceramic inks. After the transfer had been made upon the object itself, the piece was fused. A clear transparent enamel was then sifted over the design and fired again for gloss and protection. When the vogue for this type of work faded in the enamel metal field, the same reproduction process was used in the famous Wedgwood potteries.

16 Refining Ores to Make Color Oxides

ALTHOUGH the contemporary artist need not take on the role of the chemist or ceramic engineer and refine ores to make his own oxides, it is useful to know how they are made. Pure pigments are essential for the creation of any fine artwork. The substances chiefly used are oxides and other compounds of iron, chromium, cobalt, nickel, manganese and copper, together with the noble metals: platinum, silver and gold.

FERRIC OXIDE (IRON)

Ferric oxide is found in nature and occurs almost chemically pure in some minerals, e.g. iron ore and red haematite. Iron ores are very hard and costly to grind into powders. Iron oxide can be made by dissolving ferrous sulphate in water and treating the filtered solution with nitric acid, changing it to ferric sulphate. When dried, the precipitate is a reddish brown and can then be converted into pure ferric oxide by roasting it at a high temperature. Its color then changes to cherry red, purple, red-brown and violet, depending on the temperature used.

CHROMIC OXIDE

This is one of the most important glass colors because it is the only substance that yields a perfectly pure green when fired hard. One way of producing this oxide is to mix equal parts of potassium bichromate and sulphur, both very finely ground, exposing the mixture to a low red heat in a crucible. The fused mass is then crushed and washed until all soluble matter has been removed. After drying the residue will be a dark green powder consisting of pure chromium oxide.

COBALT OXIDE

This oxide is the only substance furnishing a pure blue in enamels, although vanadium-zircon compounds have been developed in recent years for giving turquoise shades. Cobalt oxide is prepared from the natural ore in which it is contained, but the separation of the other ores presents a great problem. Cobalt salts are usually used, the chief of which are the chloride and nitrate. The salt is dissolved in water and this solution is then treated with caustic soda as long as the precipitate continues to form. It is dried after repeated washings in water. The residue is then calcined to form a deep black-blue powder consisting of pure cobalt oxide.

MANGANESE DIOXIDE

This oxide is an important material for producing brown, violet and black colors. It can be prepared from a salt of manganese, preferably as free as possible from iron contaminants. The fine red crystals of this salt are dissolved in water, precipitated with caustic soda, washed, dried and calcined.

NICKEL OXIDE

This is made from nickelous sulphate obtained in the pure state as green crystals. This also must be precipitated with caustic soda, washed, dried and calcined.

COPPER OXIDE

Copper sulphate perfectly free from iron compounds is required. This salt is dissolved in hot water and treated with caustic soda until precipitation ceases. The black precipitate collected on a filter is then washed with hot water and dried.

BARIUM CHROMATE

Barium chloride is treated with potassium chromate, washed and calcined to produce a yellow pigment. Too high a temperature will give it a green tinge.

ANTIMONY OXIDE

To produce pure yellows, commercial antimony trisulphate must be reduced to a fine powder and then boiled with hydrochloric acid as long as hydrogen sulphide is given off. The liquid is then allowed to stand until quite clear and the solution drained off from the sediment. The resulting solution of antimony trichloride when poured into a large quantity of water is decomposed and settles into a dense, pure white residue. This residue is combined with other agents such as red lead to produce lead antimoniate which is known as Naples yellow, the common yellow enamel pigment.

URANIUM OXIDE

After a period of scarcity because of its strategic importance, this oxide is now back on the market in the form of spent uranium oxide. It has been used to produce beautiful yellow glass. The general tone is possibly more like a wine-yellow or yellow orange. Its coloring power is so great that one part by weight is all that is necessary to color 200 parts of glass flux a deep yellow. A range of rich warm yellow tones can be produced by the addition of small amounts of silver oxide or silver nitrate.

NAPLES YELLOW

The effective component of this color is lead antimoniate. One of the processes employed in producing it is to ignite lead nitrate with tartar emetic and sodium chloride. The color will depend on the firing temperature. Some batches contain zinc oxide. Its covering power is not very high and accurate firing temperatures are required since excessive heat will darken the stains. When this material is used as a mill addition with clear glass frits additions of 6 to 8% are needed.

CADMIUM YELLOW

Cadmium sulphide is used to produce this high toned yellow. It is obtained by precipitating cadmium salt solutions with hydrogen sulphide. Depending upon the process used the resulting color varies from bright lemon yellow to dark orange, the changes being caused by variations in the crystalline forms. Cadmium sulphide is moderately resistant to high temperatures, giving a deep intense yellow but under elevated temperatures a greenish cast may appear. It has an excellent covering power and 2% is all that is required for a mill addition.

CADMIUM RED

Cadmium red makes a bright red-orange and this is obtained by heating cadmium sulphide in the presence of sulphur and selenium. It is produced by the combined precipitation of cadmium salt and alkaline earths and selenides, following by intense heating. The resulting mixture becomes cadmium red. With clear glass frit only 2 to 3% for a mill addition is needed. With titanium glass the color will be more pastel in character.

CHROME RED

This color is due to the basic lead chromate and since lead when combined with sulphur and selenium turns black it cannot be used with selenium pigments. It can be produced by heating neutral lead chromate with sodium hydroxide. If heated above 1 550°F (845°C) chromium red decomposes. It will have a good color and gloss at approximately 1 450°F (790°C). Mill additions require about 7 to 12% for good covering power. Chrome reds are highly susceptible to crystallization on cooling, giving enamel faults such as scumming. They have largely been replaced by selenium reds in the enamelling industry, but their unique effects offer the artist-potter certain scope.

COBALT BLUE

Nearly all blue colors used in enamelling contain cobalt oxide, although in recent years new compositions based on vanadium and zirconium have been developed to give turquoise effects. Light blues are based on cobalt aluminate spinel which is very resistant to high temperatures, providing the silica content of the flux is low. Dark blues are based on cobalt silicate. Very small amounts of cobalt oxide will produce blue enamels and the oxide should be used very carefully to avoid contamination.

GREEN

Blue-greens are produced by combinations of chromium, alumina, and cobalt oxides. The enamel colors are obtained by subjecting the sulphate of these three metals to intense heat. Variations in the proportions of these oxides will yield a wide range of shades, from blue to green. By combining these oxides, the color tone becomes much more stable than with the use of a single oxide. Colors that are too intense can be toned down or made more subtle by adding opacifiers, or made softer by certain modifying agents.

If the alumina in the blue-green is replaced by zinc the resulting color tends to become greener. Chromium oxide is the basis of green colors and small amounts (3 to 5%) can be used to make green enamels. The oxide is made by heating calcium bichromate with reducing agents such as coal, sulphur, or ammonium chloride. Chromium oxide is a dirty green powder and is extremely resistant to acids. Bright colored greens can be produced by adding silica and small quantities of copper oxide. If chrome green is mixed with yellow a range of yellowish greens can be produced. By adding white, higher keyed color tones can be acquired; by using a small proportion of black, olive green can be produced.

BLACK

Intense black can be obtained by combining a range of coloring oxides such as chromium, cobalt and copper, and firing at high temperatures, and then adding this to the combination of iron oxide, nickel and manganese oxides. These oxides are not light absorbing colors in themselves, but in combination they are completely light absorbing, giving the impression of black. Mixes high in borax neutralize the various oxides' color potential and result in a more intense solid black. All blacks subjected to high temperatures will have a tendency to lose their opacity and return to bluish grey transparencies.

GREY

Grey is obtained by mixing white opacifiers with blacks. The darker the grey required the more black must be mixed with the opaque white. A warm grey is made by adding lemon to a medium grey, and a blue-grey by adding blue to a similar grey batch.

WHITE

White enamels are made by opacifying clear glasses with such agents as tin oxide, zirconium dioxide, or zircon. The amount, size and index of refraction of the particles determine the thickness of the coating required to produce a certain degree of whiteness. Zircon frits with the opacifier melted in the glass are used, but tin oxide is invariably added to the mill since chemical reduction of the tin oxide may occur, thus impairing its opacifying action.

RUBY RED

One of the most beautiful glasses of all is that made with metallic gold. Traditional craftsmen used it as a substitute for ruby, and in some instances when properly fused over copper the effect is more attractive than the color of the precious stone.

Smelting gold salts into a perfectly transparent glass yields tones varying from light reddish pink to deep oxblood red. Only minute traces of gold are needed to give a ruby glass. Several formulae are available for producing this glass, the easiest consisting of gold trichloride (obtained by dissolving gold in *aqua regia*) and using the resulting solution much diluted. This solution is sprinkled over quartz glass (or clear glass frit) in the charge, and the ingredients are well stirred to ensure that all the particles are wetted. If the glass frit itself is not used in the charge a typical raw batch for a ruby glass is as follows (by weight)

Sand	100	Saltpetre	10
Potash	34	Manganese dioxide	5
Red lead	110	Gold	0·6

The ingredients are heated in a pot crucible at high temperatures; too low a temperature will produce a brownish glass. When the molten glass is poured on to a slab of smooth steel the glass is colorless with only minute particles of ruby in evidence. When it is ground up in a powder ready for use and fired, the intense uniform color reveals itself. Ruby glass can be applied direct to Tombac or guilder's metal, producing a cherry red, but when it is used over copper a clear glass flux must first be fused to the metal as a base for the ruby. Red ruby applied over opaque white will produce a pomegranate color. Extreme care and skill is required to produce deep toned ruby or blood colors.

PINK

A satisfying pink glass can be made without using gold. The method consists in heating tin oxide, quartz and chalk with calcium bichromate and borax. Depending upon the percentage used the color obtained varies from light pink to dark violet. The maturing temperature of this color is about 1 550°F (845°C). It can be made to mature at both lower or higher temperatures by changing the proportion of the ingredients. It needs about 7 to 10% stain for good covering. With melts high in boric acid the color tone changes to a bluish or lilac shade. Calcium in the form of lime added to the glass batch will give it vibrance and life. Pink can also be made from gold salts and white combinations.

PURPLE

Purple shades with the best quality and depth of color tone are made with gold. A solution of stannic ammonium chloride is mixed with granulated tin for several days and then treated with very dilute chloride solution to produce a purple color. Adding alcohol accelerates the process. A rose purple can be made from a solution of 1 gm of gold in the form of gold trichloride, mixed with a solution of 50 gm of alum and treated with 1·5 gm of stannous chloride. Ammonia is poured in as long as the precipitate forms. The residue is washed, dried and mixed with 70 gm lead flux and 2·5 gm of silver carbonate. A rose color results.

BROWN AND TAN

A combination of bismuth, antimony and titanium is thoroughly fused; for the dark brown it is necessary to add iron oxide and ochre.

17 Creativity

LEARNING an enjoyable craft is one thing, but if the young student is to pull his weight in the world he must have courage and a real sense of responsibility. If he is to contribute something worthwhile he must have the added quality of creativity.

Intelligence alone is not a guarantee of creativity; neither is the amassing of knowledge nor the mastery of technique. Each is necessary, however, for ultimate worthwhile production. The full range of qualities that distinguishes the true creator from the outstanding technician is yet to be identified. Because the creative individual is bound to be regarded as a challenge to the *status quo*, he will need the qualities of courage, confidence and independence, as well as the strength to stand alone and resist destructive criticism. Capable art teachers are trained to recognize creative traits in young people and are doing everything possible to encourage and develop imaginative powers.

The dictionary gives several definitions of the word 'create', but one in particular stands out: 'to produce as a work of thought or imagination, especially as a work of art along new or unconventional lines'. Synonyms given are: 'make, build, produce, fashion, invent, originate'. All these definitions imply both a process and a product. Creativity at its best reflects a questioning quality. It represents the best efforts of the individual to organize processes and materials into patterns which have meaning, that is, to produce actions or objects that are in the real sense useful.

A true creation, then, should help the individual to achieve greater harmony with life. In the highest sense it will help others to find beauty and meaning in their lives and surroundings; to see, to be alert, and to understand the world about them. Possibly the most universal trait of the creative person is a willingness to work hard and long. Edison went so far as to define genius as 1% inspiration and 99% perspiration.

Certainly there is no medium in use today that challenges creativity in the person more than that illustrated in this book.

Enamel steel panels, plaques and bowls, by Thelma and Edward Winter [1959].

Lustres and liquid gold were used as accents on this panel.

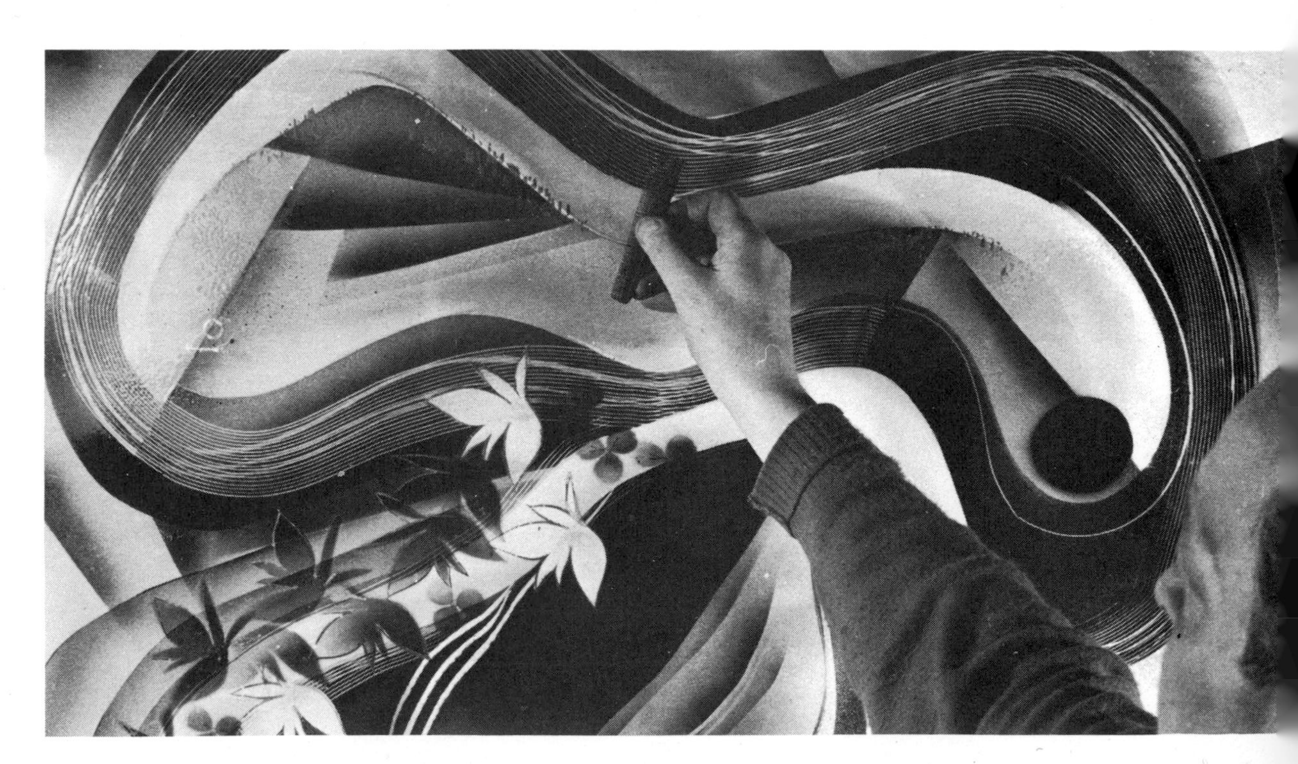

Comb proves itself as decorating tool.

Christ and the Redeemed. St Mary's Orthodox Church, Cleveland, Ohio [1960].

The Enthroned Christ. St Mary's Romanian Orthodox Church, Cleveland, Ohio [1960].

The artist shown with the 25-sectioned Enthroned Christ, 12 ft by 16 ft [1960].

PLATE 114

Enamel steel mural on exterior of athletic complex Colby College, Waterville, Maine [1968]. By Edward and Thelma Winter.

Full length view of the mural.

Mural executed in five sections of 18 gauge steel with 2-in flanged edges. Colors – brown, black, red, turquoise, grey, white and gold [1968].

Workmen hanging the ectional mural with a etal clip system.

Referring to the drawing in painting the enamel.

PLATE 116

Crown Filtration Plant, Westlake, Ohio. A 42-section enamel steel mural of water filtering processes. By Edward Winter [1956].

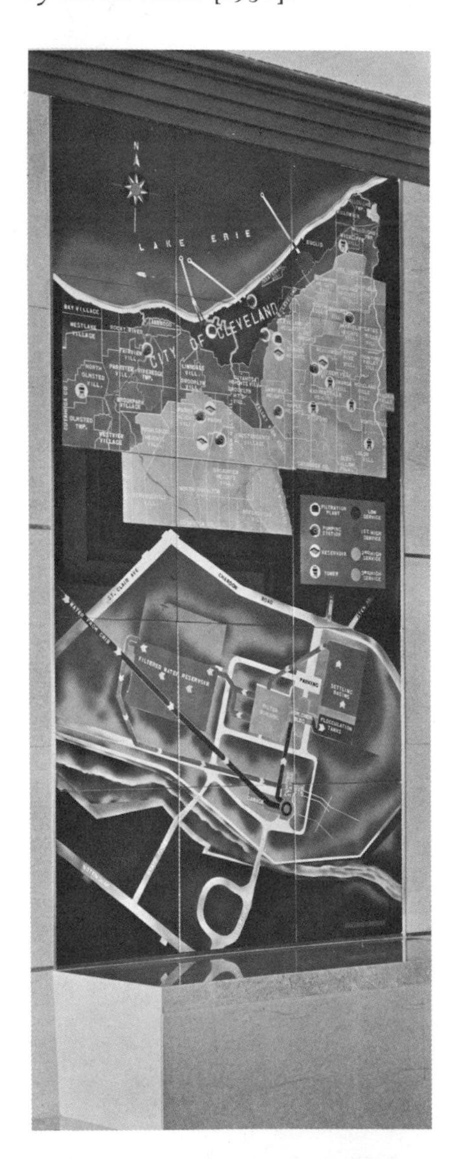

Nottingham Filtration Plant, Cleveland, Ohio. Enamel steel decoration of 18 sections [1951].

Nottingham Filtration Plant, Cleveland, Ohio. The author shown with the 18-section mural [1951].

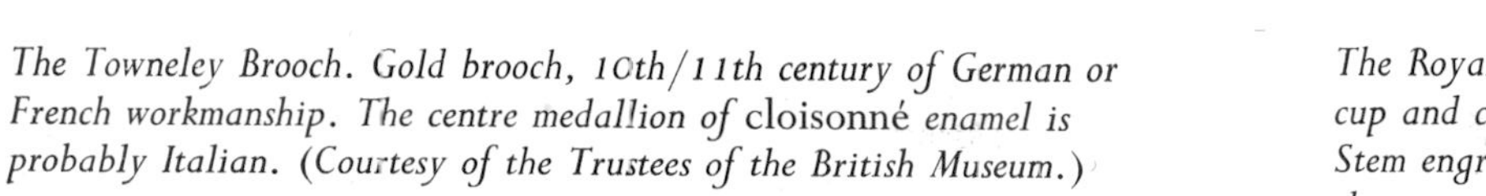

The Towneley Brooch. Gold brooch, 10th/11th century of German or French workmanship. The centre medallion of cloisonné *enamel is probably Italian. (Courtesy of the Trustees of the British Museum.)*

The Royal Gold Cup of the Kings of France and England. Solid gold cup and cover, Paris 1380(?). Footcresting enriched with pearls. Stem engraved and with Tudor roses in relief. Figures enamelled in the manner of 14th century. (Courtesy of the Trustees of the British Museum.)

Painted Enamel. French 16th Century (Limoges). Triptych; Crucification, Christ and St Veronica; Right—Pieta (Metropolitan Museum of Art, New York).

Painted Enamel, French 16th Century (Limoges). Francois I by Leonard Limousin (Metropolitan Museum of Art, New York).

Painted Enamel. French 16th Century (Limoges). St Francis by Jean de Court (Metropolitan Museum of Art, New York).

18 Enamelling – Questions and Answers

DURING the past few years the author has received many questions from amateurs and professionals. It is satisfying to tackle technical problems that puzzle individuals, since by so doing we can win many friends, knowing that kindness and advice may help the amateur to make the transition to professional status. It is hoped that the following practical problems and the answers made by the author will be of assistance to the newcomer to enamelling.

Running Colors

QUESTION: After painting the color on the white fired surface I tried to scratch lines and draw into it; within a minute the liquid colors flowed together and disfigured my drawing. What was I doing wrong?

ANSWER: After you have applied the oil colors to the surface the piece should be placed in a warm place, in a stove, or over a hot plate partly to dry. This may take only 20 min until the surface can take a sharp pointed tool and retain the line. If the color is applied too heavily and dries too hard the drawn line will chip off in places and be irregular. Dental tools, sharp knives, or any hard stick with a sharp point will produce the line desired. Practice with all types of tools will produce the desired effects.

Copper Engravings

Q: I have been an etcher for several years. Is there any possibility of using my copper engravings as a base for transparent and opaque enamels?

A: Yes. When glass enamels are fused over etched copper you are working in the traditional process known as *champlevé*. Engraving plates are heavy and will require a higher temperature than is normal with 18 gauge metal. This technique has many possibilities for the artist. Try etching the design twice as deep as usual; this will give enough depth in the metal to apply an opaque white or black powder enamel ground as fine as 250 mesh into the engraving; and then fire a transparent clear enamel over the entire metal surface. A range of enamel colors can be used to vary the effect. A 12 or 14 gauge piece of copper should be fired for about 4 min at 1 500°F (815°C).

Economics of Enamelling

Q: Is it possible for a person to make a good living making enamels? Most artists I know graduate from art school or university and commence teaching. Could they do better working at their art or craft eight hours a day?

A: A book could be written about the artist making a living with his art. Some artists are not persistent enough and fail to find markets for their work. A person must be dedicated to his work, have unbounded energy, drive, and a willingness to work hard. He must have utmost faith in himself, be willing to lay out money for equipment, tools and materials, and set a daily work schedule for himself. Although he is his own boss, he soon finds out that when he is not working and producing he is not eating or paying his bills. For sustained success one must start out in a small way: discover a customer who is interested in fine work and interested in patronizing you. Most beginning craftsmen make the mistake of placing their fine work on a shop shelf on a consignment basis; this may be good for the shop owner but not for the craftsman. If the artist plans an intelligent line of items (various shapes, sizes and colors) at a varied price range he should make the prices attractive enough so that the store buyer can purchase a large quantity outright. This realistic pricing will then give the store the opportunity of marking the work up in line with all other products on the market. An outright sale of 20 or 30 items will make a good display and give the craftsman ready cash with which to purchase more materials and continue his production.

To win success in any field of endeavour one must learn good human relations and be ever ready to co-operate with anyone who shows an interest in you and your art. If the merchant can make a living from your energy and talent he will value you and do his best to promote and sell your work.

To survive in our complex society on the sole output of your art requires many things not covered in this short answer. Of all the attributes required I would say hard work tops the list: this combined with skill, good design and color sense, craftsmanship and the willingness to co-operate with others will produce success.

Metal-Glass Reactions

Q: Will you explain the reaction between metal and glass enamel during firing? Do all metals react in the same way?

A: This subject is very complex and entire books have been written on it. Firing is a very important process in enamelling and many complicated physical and chemical reactions take place in the process. Simply, what happens is that the glass enamel fuses and reacts with the underlying metal, thus bonding the two layers together. The type of bonding and the reactions depend very much on the type of metal and on the chemical composition of the enamels. Certain agents are incorporated in enamel fluxes to aid bonding and to produce strong enamel-metal systems. For instance, the bonding of enamel with platinum, gold and silver is due to mechanical effects. A thin film of oxide on the metal surface guarantees that it is wetted by the enamel melt and provides contact between the glass and the metal. In order to obtain a strong bond the surfaces of these noble metals must be roughened chemically or mechanically. The enamel melt fills in the depressions in the surface and holds on to them after the enamel solidifies. The enamelling of copper involves bonding on a smooth surface. This is done because of the layer of copper oxide formed on the boundary between copper and enamel. The copper oxide is dissolved in the enamel on one face, and on the other diffuses into the crystals of the metallic copper, forming a strong bonding layer. When iron is being enamelled, rough surfaces can be made even rougher by sandblasting to aid bonding. The enamel melts and flows into the depressions giving strong bonds. Iron also oxidizes during enamel firing and the layer of oxide increases in thickness as the glass reacts with the material, thus partly dissolving the oxides. Bonding in this case is different from bonding with copper since solution of iron oxide does not occur in iron. The nature of bonding between aluminium and enamels is particularly complicated and research is still going on into this subject. Apparently bonding occurs through an intermediate layer of aluminium oxide. Indeed, research into the precise nature of all types of metal-enamel bonding is continuing, and even for enamel-steel bonding there is no single opinion about the bonding mechanism.

Saving Spoiled Ware

Q: I have produced a copper enamel ashtray and a small bowl but the results were unsatisfactory. Is there any way of saving these?

A: An enamellist can cover up a bad design by applying opaque enamel. There is, however, a limit to the thickness of the application before it may pop off in cooling.

I suggest placing the piece over a steel stake and pounding off the enamel. After reforming the shape start again with the application of enamel. This can also be done with steel based enamels. Protect the eyes against flying glass.

Chrome Plating

Q: Is there any way of enamelling a section of a design on steel, and chromium plating the bare section of metal?

A: This can easily be done. The paper template or masking tape should be securely fixed to the bare metal so that none of the enamel spills on the metal. This technique is not common but presents great potential.

Spraying with Airbrush

Q: Is it possible to spray enamels with a small hand airbrush? The enamels I use are ground dry at 80 mesh and must be applied with a hand sieve.

A: All types of enamels (opaque or transparent) can be ordered from suppliers for spraying with a hand gun or airbrush. Any commercial illustrator accustomed to an airbrush will find vitreous enamels behaving similar to water paints or inks in a Binks Wren airbrush. Workers used to an airbrush will find a little practice necessary to obtain the necessary gradation of color tone, but it is not difficult. When ordering enamels ask for enamels for airbrush spraying. These are finely ground and in some cases alcohol is used as the suspension agent. Both transparent and opaque enamels sprayed over a fired opaque white surface will give the best results. Spraying directly on copper surfaces will produce a very thin color. Several coats and firings may be necessary to obtain deep transparent colors directly on copper. Traditional enamellists used pointers and spreaders to apply the 80 or 90 mesh enamels to metals. Fine grinding and spraying was unheard of even thirty-five years ago. The enamellist of today is very fortunate indeed.

Cracking in Refiring

Q: In my first firings of transparent enamels on copper I get a beautiful thin transparent surface. When I refire them after applying enamels to the back side they have a tendency to crack in the second firing and the quality is duller and less transparent.

A: The kiln temperature is probably too high, causing the enamels to run and thin. Fire at 1500°F (815°C) or slightly lower for 2 min, then sift on a second light coat and apply enamel to the back side in firing for the second time. The second or third firing can be done at a good red-whitish heat and will give you the desired depth of color. A gradual building up of temperature is better than overfiring first time. The proper feel of firing can only come through experience and many firings.

Gold and Lustres

Q: I enjoy using the liquid bright gold on enamel surfaces but I am told that any number of colored lustres can also be made with gold. How is this accomplished?

A: Some of the most beautiful lustre effects can be produced with bright gold in combination with bismuth lustre. For instance, 5 parts of bright gold and 1 part of bismuth lustre fired together on an enamel surface will give a coppery sheen. If the proportion is changed to 2 to 3 parts of bismuth lustre to 1 part of gold, bluish violet colors with a peculiar gold sheen result. Two parts of bright gold and 1 part of bismuth lustre will give a rose red. One part of bright gold to 4 parts of bismuth lustre produces a pale blue. These effects are due to the various states of divisions of the particles of gold. It is true that gold is capable of giving all colors in the spectrum, depending on its state of division (ranging from molecular sizes to a few microns). These colors can be further modified by adding certain amounts of other lustres to the gold-bismuth combinations, e.g., iron, chromium, uranium, etc. Some European producers make attractive lustres by combining bright gold and sulphur balsam. An imaginative artist need not be a chemist to obtain interesting and delightful effects with metallic lustres. By purchasing small quantities of these lustres one can create new colors and precious looking tones by firing them on to enamels. Lustre colors of all types will appear more true in tone when fired over a titanium coated steel base rather than sheet copper, because copper yields copper oxide fumes when heated, and this turns lustre colors greenish, thus spoiling the desired shade. Steel remains fairly resistant and does not affect the color coating or the lustre fired over it.

Opal Effects

Q: I would like to produce a semitransparent colored surface that would give the appearance of an opal. Is this possible with enamel?

A: Enamel manufacturers make opalescent enamels that fire over copper into a wonderful milky semi-opaque transparent surface. These assorted colors must be fired very carefully since too long a firing will kill the transparency and make the surface opaque. The enamel is made with a transparent base and includes small additions of opacifier and coloring oxide. When a beautiful surface appears do not refire, since a second firing will spoil it. An opalescent bowl or ashtray becomes a collector's item, as the quality of running glaze is quite unique.

Protective Glasses

Q: Is it necessary to wear protective glasses when firing enamels?

A: Anyone working continuously at the open door of a muffle kiln (enamels fired in 3 min) should wear protective glasses. The best type contains blue glass that cuts off the infra-red radiation from the open kiln. For occasional firings in small studio kilns it is not necessary to use such glasses, but for fast production or continuous firing they are essential. Even plain glass serves to protect the eyes from the heat. My own eyesight has been damaged slightly by working without glasses for many years in front of a huge enamelling furnace. However, for the past twelve years I have found glasses excellent protection and now I would not work without them.

Irregular Application

Q: I have difficulty in producing a smooth, clear transparent color after firing. The enamel seems too heavy and irregular. Is my application wrong? I use a heavy application of gum tragacanth.

A: Many enamellers think gum tragacanth should be handled and applied like glue.

Proper use of this vegetable gum involves thinning it with water, boiling it in a basin and filling the pan with water and only a few flakes of the whitish gum. After slow boiling for an hour it can be strained through fine cheesecloth. When gum is applied with a brush to copper it should look almost as clear as water. After the powdered enamel has been screened on to the piece through an 80 mesh screen a second application of thin gum can be sprayed on, using a fixative mouth sprayer.

Speckled Surfaces

Q: I have been trying to obtain speckled textures but all I get is a dusty surface with tiny speckles. I use hard opaque white. Where am I going wrong?

A: Your application of opaque white is too thin. Large speckled surfaces can be obtained by applying the white enamel heavily, with almost any type of transparent color being applied on the top. Fire at 1 480° to 1 500°F (805° to 815°C) for about 3 min. It is more difficult to produce large white speckles on a flat piece of metal, but it is possible by resting the piece more vertically on the firing trivet. By allowing the enamels to run in the firing a large speckled texture surface can be produced. Too much firing causes the speckling to melt and gives a runny, streaked surface. Proper application, drying, firing and timing is essential to obtain this effect.

Stainless Steel Enamelling

Q: Can stainless steel be enamelled successfully?

A: Most opaque steel enamels can be fired successfully on to stainless steel but the surface to be coated must be sandblasted to obtain a dull surface before the enamel will adhere.

Opalescent Enamels

Q: What are opalescent enamels and how do they differ from the opaque or transparent variety?

A: Opalescent glass enamels are made with a transparent base but are compounded with an opacifier. In lump form they look more opaque than transparent. However, when powdered or finely ground and applied to copper followed by firing at about 1 500°F (815°C) for 3 min, the enamel turns semitransparent and flows over the sides of a copper bowl in a rich milky effect. Temperatures and timings are important since a second or third firing will probably turn the semitransparent surface solid opaque. In working with opalescent enamels I advise the enameller to value and appreciate a beautiful effect when it is obtained; trying to improve it by refiring may end in disappointment. Some color tones, and running rather unique effects, may never appear again. Many opaque jewellers' enamels will produce a semitransparent surface when overfired. Keeping a notebook with color numbers, firing temperatures and timings is invaluable. Too few artists working with enamels make enough test samples. Any conscientious and imaginative worker should have 60 to 70 color samples and learn how colors behave under fire before attempting to make large works of any kind.

Quartz Lamps

Q: Have you had any experience with quartz lamps incorporated in a small kiln for the firing of enamels?

A: There is an infra-red kiln on the market in the USA but I have not tried it for firing enamels. The lamps fuse the enamels from the topside down. Since heat rises I believe the best enamelling results will come from putting the heating elements on the bottom or on the lower sides of the kiln. When intense heat passes through and permeates the metal and enamel, the glass flows quickly and produces a beautiful color and surface.

Lumps or Powders

Q: In purchasing enamels, do I buy the enamel in the lump form or as powders, finely ground?

A: If you intend using only small quantities of enamel it is best to order from the supplier in an 80 mesh grind. This is then ready to be screened through an 80 mesh sieve. Finely ground enamels stored for long periods tend to lose their clear transparency, or may fire with pitted effects. By using a mortar and pestle (steel), or a crusher, the enameller may find it convenient to order lump frit and then grind to 80 mesh. A small charge is made by the supplier for grinding enamels but most small-scale enamellers purchase it ground ready for use.

Enamelling Noble Metals

Q: Please give me some details about enamelling silver and gold.

A: Pure silver and gold are expensive and for this reason are not widely used as are copper, steel and aluminium. The application of enamels to noble metals is generally the same as for copper: cleaning the metal surface, then wetting it with a brush and gum tragacanth. Enamels are applied by sifting through a screen. Enamels designed for silver are harder and fire out in a higher color key. They should be ordered specifically for silver. Any transparent copper or silver enamel can be fused to cast on sheet gold. Firing temperatures should be 1 500° to 1 520°F (815° to 830°C) (a whitish red heat). When enamelling gold it is best to leave some of the metal bare with the transparent enamels embellishing other areas. Since silver tends to tarnish when exposed to the atmosphere (due to reaction with sulphur, forming silver sulphide) it is a good policy to cover all silver surfaces with transparent enamel. This will eliminate the need for rebuffing any bare metal areas that tarnish.

Which Side First?

Q: Do you enamel one side of a copper ashtray, dry and fire, and then clean the back side, enamel and fire; or should it be done in one operation? When I fire a piece the enamel tends to pop off when it hits the heat of the kiln. What am I doing wrong? After an ashtray is fired should the copper edge be polished or left as it is?

A: While it takes a little more patience and skill to apply enamel to both sides of a tray and fire at the same time, this is the method we were taught at the Kunstgewerbeschüle in Vienna. If the inside is enamelled leaving the back side bare, the copper will scale and oxidize. When the piece cools this must be cleaned off with sulphuric acid and water (5 parts water : 1 acid). The powdered enamel is then applied on the piece and fired inverted. If your enamel pops off when firing it means it has been applied too thickly, with the use of too much gum; the fault may also be due to insufficient drying. The work must be absolutley dry before it is inserted into the kiln. The edges of all pieces should be stoned (using a carborundum wheel), rubbed with fine emery paper and then buffed on a motor wheel using rouge. A true craftsman believes in technical excellence. A beautifully polished edge on the copper or steel will result in greater customer acceptance and a better chance of a sale. Enamels and metals are a jeweller's medium and all fine jewellery must have fine polished surfaces. Metal edges can be protected after polishing by coating with hard Egyptian lacquer on a camel hair brush.

Speckled Fault

Q: When I apply opaque black enamel to copper and fire it I get a speckled surface. I wish to obtain an even glossy surface. What am I doing wrong? This happens with opaque white and dark blue also.

A: If any opaque colors speckle during firing it usually means that the temperature is too high. You are using a white red heat instead of a low to medium red heat. Set your pyrometer at 1 450°F (790°C) and not 1 520°F (830°C). Interesting speckled and flow textures can be obtained by using elevated temperatures or leaving the piece in the kiln for 5 min instead of 3. Technical efficiency is obtained by controlling the temperature as well as the firing time.

Copper Versus Steel

Q: What, if any, are the advantages of steel over copper? Is a particular type of steel necessary?

A: Steel has some advantages over copper. It is stronger, will withstand more abuse than copper, and serves well in decorative accessories for the home and in architecture. In using steel as a base metal one must employ opaque enamels in coating the surface of the metal. After firing, a variety of enamels can be applied. This will give a high keyed color and not the depth of transparency obtained in applying the transparent glass enamels directly over copper. Many design and texture possibilities are available with the use of steel enamels. Line design, sgraffito, white-line drawing, lump textures, and many other surfaces are possible. Steel is cheaper than copper. Low-carbon steel should be used. A dark blue ground coat enamel is applied to the steel as a first coat before the white and transparent colors are applied. Metal and steel enamels can be purchased from suppliers. Clean 18 gauge steel will need to be cleaned with scouring powder, brush and water.

Natural Copper Color

Q: I wish to enamel copper sheets 30 x 26 in in size with transparent enamel so that the natural color of the copper shows through. Is it possible to enamel over copper plated steel to save the expense of buying copper sheet? In firing liquid gold on copper enamels I notice that when the gold has fired properly the enamel begins to crack. Further firing to heal the cracks makes the gold split. Can you suggest a remedy?

A: The only manner in which transparent clear enamels can be fused on to sheets of copper without showing mottled, spotted or cloudy surfaces is by spraying. Transparent clear enamel can be milled in alcohol, and then by lightly spraying on to a clean copper surface it is possible to produce an even coating showing the true color of the copper. It is possible to enamel over copper plated steel providing the enamel is a low melting type, that is, one fusing at 1 200° to 1 300°F (650° to 705°C). Copper plating will burn up at 1 500°F (815°C). This problem is worth experiment.

Copper enamels fired over copper have a wide range of expansion and contraction. I suggest cutting down on the temperatures in the kiln and heating the pieces gradually. The higher the temperature and the longer the piece is in the kiln the more it will crack. The enamel surface should be fused well at 1 400° to 1 480°F (770° to 805°C), and when the gold is being fired the temperature should be reduced to 1 250° to 1 300°F (670° to 710°C). Make the piece thoroughly warm before placing it in the kiln. Two minutes at the above temperature should be sufficient to fuse the gold. A good smooth bright gold surface without cracks can only be obtained over hard steel enamels and these should be fired at about 1 300°F (710°C). Burning away the volatiles before the piece is placed in the kiln is essential. Working with 18 gauge copper will give less trouble than with lighter metals.

Fondant

Q: Please explain the meaning of the word fondant.

A: Fondant is another name for clear flux and is used in France, Italy, Germany and Austria. Flux is the English term meaning clear glass.

Enamelling Platinum

Q: Can platinum be enamelled? If so, do you need special types of enamels, and how is the metal prepared before enamelling?

A: Platinum is rarely used for enamelling because of its high price. But as a metal for enamelling it presents hazards. The matching of metal and enamel with respect to expansion is very difficult. The surface of the metal is very difficult to enamel; the coat tends to pop off during cooling. Better adherence can be obtained when the metal is gouged by hand or machine-tooled with lines or designs that have a tendency to hold enamels to the surface. During the firing of regular transparent enamels on platinum, try cooling the piece gradually, and do not bring the piece from the kiln too quickly. A gradual cooling can take place at the door of the kiln. Fast cooling, as with copper, silver or steel will cause cracking and possibly shearing off of the enamel in the cooling.

Pottery Kilns for Enamelling

Q: Is it possible to fire enamelled copper in a pottery kiln?

A: Enamelling can be done in pottery kilns, but an easily removable shield acting as the door should be devised for quick removal of the enamelled articles. This shield should be replaced quickly to avoid heat losses. Various types of small potter's kiln are available in America and Britain.

Enamelling on Gold

Q: With me enamelling is just a hobby, like panning for gold in South Dakota. I work at it in the evenings and during weekends. My furnace measures 6 x 6 x 12 in, which allows me plenty of freedom. I never took a lesson; just taught myself from your books. I price my work rather high – from 200 to 2 000 dollars for an ashtray. A jewellery shop in Homewood, Illinois, features my work. Having been a gold prospector I fuse various sized gold nuggets into the transparent enamels. My 2 000 dollar ashtray contains large hunks of gold. Can you suggest a soft fusing clear enamel which would melt and give me suitable pools with which to sink my gold nuggets?

A: A No. 426 clear flux enamel from well known suppliers might be useful here. However, any transparent color No. 122 blue, No. 111 light blue, or No. 110 sapphire blue, for instance, should melt well into deep pools of enamel. Apply plenty of enamel and give the heavy gauge copper a high fusing of about 1 520°F (314°C) for 4 to 5 min. This will give the gold nuggets a beautiful pool in which to lodge. The blue colors complement the warm gold nugget and then you can perhaps raise the price of an ashtray to 3 000 dollars! Just kidding!

Spun Metal Shapes

Q: My friends who enamel copper use spun metal shapes, including ashtrays, bowls, etc. None of their work is unique or seems to have any character about it. Your comments on this would be interesting.

A: Enamellists who limit themselves to enamelling prespun shapes are taking the easy way out. Most of the shapes are not well designed and much of the time, material and effort going into these pieces is wasted. Occasionally an object made by a skilled worker will be very beautiful, the quality of color and transparency being unique, but the amateur will usually apply the enamel too heavily and the results will be disastrous. The professional metal craftsman knows how to make his own copper shapes and in so doing is able to planish (hammer) the surface of the metal. This planished surface, catching the reflected light from each hammer mark, produces a depth of color found in no other material. Hand hammered works are much more valuable than spun pieces and bring a higher price. Many craftsmen take the easy way out but in so doing they cease to grow in technical excellence.

Special Copper Effects

Q: Is there some way of getting effects on copper other than the transparent enamels fused on the smooth metal?

A: Sheet copper can be rubbed with emery paper, sandpaper, scratched with sharp tools, or hammered; pouncing designs can be obtained with engraving tools, and punch and hammer designs can be obtained. The craftsman will find that smooth and textured surfaces in combination will give greater contrast and interest to the metal which is to be enamelled.

Spraying Aluminium

Q: Can glass enamels for sheet or cast aluminium be sprayed? If this is possible how heavily are the enamels applied? Should the enamel be dried out thoroughly before fusing?

A: Glass enamels compounded for fusion on to aluminium are ground so that all the glass passes through a 325 mesh screen. This degree of fineness facilitates spraying through almost any sprayer except the very fine airbrush designed for liquids only. Some producers supply large airbrushes capable of spraying finely milled glass. The Jetpak aerosol hand sprayer also works well. Transparent enamels for aluminium are best sprayed with thin coatings. Too heavy enamel concentrations make the glass more opaque. These liquid enamels can also be applied with eyedroppers or rubber ball syringes. It is not usually necessary to dry out before firing the piece (as with steel and copper enamels). The heat of the kiln or hot plate will dry the enamel before fusion commences. Small pieces of aluminium can be fused by placing the piece on the ribbon burner of an electric kitchen stove adjusted to a medium heat.

A 10 min period is required to fuse this glass to the metal. Most items are more decorative when parts of the enamel metal are left bare. This effect contrasts well with the colored enamel in the design. The bare aluminium can be polished with fine steel wool. Finished work looks like enamel on silver.

Transparent Copper Enamels on Aluminium

Q: Can we use transparent copper enamels for aluminium? What is the best way to enamel aluminium to get the highest quality results?

A: Since aluminium melts at a lower temperature than copper, entirely new types of glass have to be formulated. These glass enamels are compounded to fuse at 1 000°F (537°C) in 10 min. Aluminium will melt at 1 200°F (650°C) and copper and steel enamels are made to fuse at 1 350° to 1 600°F (730° to 870°C). The degree of transparency of some enamels on aluminium depends on the evenness of spraying the coat from the aerosol jet sprayer or any popular airbrush and compressor. For small pieces of jewellery, a trailing effect of enamel applied with an eyedropper is adequate. The best quality with this medium will come from leaving areas of the aluminium plain, and after the enamel has been fused it can be polished with fine steel wool or a buffer. This produces a fine contrast of color and aluminium which gives the effect of fine silver. I find the simple range of colors such as turquoise, black, blue, green, red and white to be most effective. Raised textures are possible by applying small lump enamels and fusing them so that they stand out like little balls on the surface of the metal.

19 Some Useful Tables

TABLE 1: COMPARATIVE TEMPERATURES

Temperatures		*Orton cone numbers*	*Color inside kiln*	*Firing range of clays glazes, and colors*	*Melting points of metals*
1 085° F		022	First	Glass golds	Tin 450°F
1 103		021	glow		Cadmium 610
1 157		020			Lead 621
					Zinc 787
1 166		019			
1 238		018	Dull red	Overglazes	Aluminium 1 216
1 328	WHITE CONES	017			
1 355		016			
1 418		015		Pottery golds and lustres;	
1 463		014		Very soft glazes; Medium	
1 517		013	Red	enamels; Hard enamels	
1 544		012			
1 607		011			
1 634		010			
1 706		09		Earthenware bricks and	Silver 1 762
1 733		08		flowerpots; Soft glazes;	Gold 1 945
1 787		07		Tin glazes; Majolicas	Copper 1 980
1 841		06	Cherry		
1 886		05			
1 922		04			
1 976	RED CONES	03			
2 003		02			
2 030		01		Hard glazes; Stoneware;	
2 057		1		English porcelains	
2 075		2	Orange		
2 093		3			
2 129		4			
2 156		5			
2 174		6			
2 210		7	Yellow	Salt glaze	
2 237		8			
2 282		9			
2 300		10	White	Japanese porcelain	Nichrome 2 550
2 345		11			
2 390	WHITE CONES	12		Chinese, German and	
2 462		13	Brilliant	Sèvres porcelain	
2 534		14	white		
2 570		15			
2 642		16	Blue	Copenhagen porcelain	Nickel 2 646
2 669		17	white		
2 705		18	Dazzling		
2 759		19	white	Spark plugs	Iron 2 741
2 786		20			

TABLE 2: COLOR TEMPERATURES

Judging Temperature by Color

°F	°C	Temper color
380–400	200±	Pale yellow
420–440	220±	Straw yellow
460–480	240±	Yellowish brown
500–540	270±	Bluish purple
540–560	285±	Violet
560–580	300±	Pale blue
600–640	325±	Blue – visible color
1 000	540	Black
1 100	590	Faint dark red
1 200	650	Cherry red (dark)
1 300	700	Cherry red (med)
1 400	760	Red
1 500	815	Light red
1 600	870	Reddish orange
1 700	930	Orange –
1 800	980	changes
1 900	1 040	to –
2 000	1 090	Pale orange lemon
2 100	1 150	Lemon
2 200	1 205	Light lemon
2 300	1 260	Yellow
2 400	1 315	Light yellow
2 500	1 370	Yellowish grey: 'white'

NOTE: The colors are for medium daylight. 'Color temperatures' are useful as a rough guide though with practice surprising accuracy can be secured as long as the conditions are held constant.

TABLE 3: THE MELTING POINTS OF METALS

Metal	*Centigrade*	*Fahrenheit*
Aluminium	659·7	1 219·6
Brass	1 015 (approx.)	1 859 (approx.)
Bronze	1 020	1 868
Copper	1 083	1 981·4
Gold (24 carat)	1 063	1 945·4
Gold (18 carat)	927	1 700
Gilding metal	1 065	1 950
Cast iron	1 100 (approx.)	2 012 (approx.)
Iron (pure)	1 535	2 795
Lead	327·4	621·32
Platinum	1 773·5	3 192·3
Silver (fine)	960·5	1 728·9
Silver (sterling)	898	1 640
Steel	1 350 (approx.)	2 430 (approx.)
Tin	231·8	417·5
Zinc	419·4	755

TABLE 4: THE BROWN AND SHARPE (B. & S.) GAUGE FOR SHEET METAL

Gauge number	*Thickness*	
	in inches	*in millimetres*
3/0	0·409	10·388
2/0	0·364	9·24
1/0	0·324	8·23
1	0·289	7·338
2	0·257	6·527
3	0·229	5·808
4	0·204	5·18
5	0·181	4·59
6	0·162	4·11
7	0·144	3·66
8	0·128	3·24
9	0·114	2·89
10	0·101	2·565
11	0·090	2·28
12	0·080	2·03
13	0·071	1·79
14	0·064	1·625
15	0·057	1·447
16	0·050	1·27
17	0·045	1·14
18	0·040	1·016
19	0·035	0·889
20	0·031	0·787
21	0·028	0·711
22	0·025	0·635
23	0·022	0·558
24	0·020	0·508
25	0·017	0·431
26	0·015	0·381
27	0·014	0·376
28	0·012	0·304
29	0·011	0·29
30	0·01	0·254
31	0·008	0·203
32	0·0079	0·199
33	0·007	0·177
34	0·006	0·152
35	0·0055	0·142
36	0·005	0·127

TABLE 5: METAL ALLOY COMPOSITIONS

Metal	*Composition*
Brass	7 parts copper, 3 parts zinc
Bronze	95 parts copper, 4 parts tin, 1 part zinc
Gilding metal	5 parts copper, 1 part zinc
Gold (18 carat)	36 parts gold, 7 parts silver, 5 parts copper
Silver (sterling)	925 parts silver, 75 parts copper
Silver solder (hard)	3 parts silver, 1 part brass wire
Soft solder	2 parts tin, 1 part lead

20 Glossary

ACID: chemical compound used to clean, etch, or pickle metals; sulphuric or nitric acid with water is used for steel or copper.

AIRBRUSH: a manual instrument used to apply enamels and lustres.

ARSENIC: a highly toxic metallic element sometimes used to enhance the gloss of transparent enamels.

AQUA REGIA: a mixture of hydrochloric acid and nitric acid which dissolves gold and other noble metals.

BALL MILL: a rotating cylindrical grinding mill in which enamel frit is wet or dry ground. The liquid produced in wet grinding is known as slush or slip enamel. Ball mills may be made of steel or porcelain. They are available in a wide range of sizes. Sometimes rubber mills are used when it is essential not to contaminate the charge.

BUFFING WHEEL: a high speed electric shaft upon which cloth, felt, carborundum or rubber wheels can be attached to polish metals.

CALCIUM SULPHIDE: a compound used in fluorescent pigments.

COUNTER ENAMEL: the enamel that is applied and fused to the reverse side of the metal. It keeps the metal from warping which would result if only one side were enamelled, leading to differences in the coefficients of thermal expansion across the article.

COMPASS: a device to produce circles on paper or for applying precious metals to enamel surfaces.

CRACKLE: may be a fault in the form of cracks in the enamel, or a deliberately designed decorative effect.

CRUSHER: a steel machine in which raw materials and lump enamels are reduced to a powder.

DETERGENT: water-soluble cleaning agent. Hot detergents are used to clean grease and dirt from metal surfaces.

DRAWING PENCIL: a specially compounded pencil for drawing white, black and colored lines on black or white matt-fired enamel surfaces. Will not burn away in furnace heat.

DRY PROCESS: the technique of sifting enamel powder on to metal using metal or fabric sieves.

DUSTING: applying dry enamel in the extremely finely ground state to an enamel surface.

EMERY CLOTH: an impregnated cloth of fine or coarse emery for smoothing metal surfaces, or the edges of enamel objects.

ENAMELLING IRON: a steel which is extremely low in carbon, made especially for enamelling. Sold in sheets or rolls.

ETCHING: using sulphuric or nitric acids with water to produce designs on metals; or the use of alkalis or corrosive chemicals for the cleaning of aluminium or steel prior to enamelling.

FELDSPAR: a mineral containing potash, soda silicates and mica; used in ceramics and glass as a flux.

FIRING: a term used for fusing enamels in a furnace.

FRIT: small particles of enamel produced when molten enamel is poured from a smelter into a tank of water.

FURNACE: a gas or electric fired chamber or muffle used for firing enamel on to metal. A 'kiln' is used for the firing of claywares.

GAUGE: an index number used to denote the thickness of sheet metal.

GLASS BEADS AND BALLS: small particles of glass used to develop special textures when fired on to an enamel surface.

GLASS STRINGS: threads of strings of drawn-out glass enamel, made by pulling molten enamel with an iron rod, and allowing it to cool. Used for textural effects.

GUM TRAGACANTH: a vegetable (seaweed) gum in hard flake form. A few flakes boiled in water produce a thin adhesive for wetting the copper for dry-process application of enamel.

HAMMER MARKS: an even pattern of marks by hammering metal over a steel stake with a planishing hammer. A surface visible through the fired transparent enamel (copper, silver or aluminium).

INORGANIC: applied to many substances that do not contain carbon as a constituent, e.g. metals, rocks, minerals, enamels and a variety of earths. Organic materials can be combusted; inorganic cannot.

LAVENDER OIL: a colorless, essential oil distilled from lavender flowers which is used in some enamelling media.

LEADBEARING: a term specifically applied to enamels in which lead oxide is used as one of the principal fluxes. It brings the melting temperature down.

LUMINOUS PIGMENTS: luminescent, fluorescent and phosphorescent materials that activate (glow) when subjected to ultra-violet or black light (radium materials).

LEADLESS: any enamel which does not contain lead.

LUSTRE: a pearly thin coating of a metallic solution fused on to the enamel surface. Made from silver, gold or copper.

MATT ENAMELS: a dull or 'no-gloss' enamel upon which one can draw and paint.

MELTING POINT: the temperature at which a solid changes into a liquid.

MESH: the numerous openings of a screen or sieve. A 200 mesh sieve has 200 openings to the square inch, an 80 mesh sieve has 80 openings to the square inch.

OILS: organic liquids in which enamels are finely ground to make a paste suitable for silkscreen application or painting.

OPALESCENT ENAMELS: enamels having a milky appearance; semi-opaque and transparent.

ORGANDY: a thin material similar to silk used for silkscreening designs on to enamel surfaces.

OPAQUE: an enamel which is not transparent to light. It covers metals so that the texture of the surfaces are completely hidden.

PAINTER'S ENAMEL: a transparent base high-lead enamel prepared by grinding in alcohol or oil; available in all colors and values.

PICKLING: the practice of cleaning and treating metals, preparatory to applying the enamel, by dipping in a hot acid-and-water bath.

PLANCHES: firing support for enamel.

PLANISHING: a hammer-marked surface on metal made by hammering metal over a steel stake.

PLATING: an electroplating process whereby one metal is coated with a thin layer of another metal, such as silver or aluminium plated steel.

PLATINUM: a silver liquid grey metal soluble in *aqua regia*. Can be purchased in liquid, leaf or paste form. Used for decorating fired enamel surfaces.

POT SMELTER: a high fired clay crucible in which glass enamels are melted.

PYROMETER: an instrument for measuring the degree of heat in an enamelling furnace.

RARE EARTHS: minerals of a precious and semi-precious nature.

RESINS: gums made from balsam trees, used in painting enamels.

RESPIRATOR: a mask to cover the nose and mouth when working with acids, lead enamels, etc.

ROLLER: cylinder with handle which can be used to apply liquid enamels.

ROUGE: a cake of red compound applied to a cloth buffer to put a high polish on metals.

RUBBER MASKOID: liquid cement to produce design and texture patterns.

RUBBER SQUEEGEE: a narrow band of hard black rubber encased in a piece of wood used to force enamel paste through silkscreen stencil.

SGRAFFITO: an ancient pottery decorating techinque, adapted by the author for use in enamelling. Effective for producing line drawings on enamel surface.

SLIP: finely ground liquid 'slush' produced by grinding in a porcelain ball mill with water and clay as suspending agents.

SLUSH: same as 'slip' for dipping or spraying enamel on to metal or fired enamel surfaces. Water and clay used as a suspending agent. Will produce crackle when fired over transparent surface.

SPOONING: applying 'slip' or 'slush' enamel with a spoon or other small dispenser. Similar to slip trailing with rubber syringe applicator.

SPATULA: a thin metal blade with a handle used to mix enamel screening pastes on a slab. For applying enamel to a fired enamel surface.

SOLDER: a fusing metal or alloy used to unite two or more metals under heat. Silver solders (hard) are used for metals that are to be enamelled at temperatures of 1 400° to 1 500°F (770° to 815°C). Soft solders cannot be used for enamelling.

SPRAY GUN: larger than airbrush, used to apply enamels.

SQUEEGEE OIL: oil used in grinding enamel for silkscreen printing.

STAKE: round, highly polished metal forms upon which metals are hammered, producing an even planished surface (copper, silver or aluminium).

STEEL WOOL: usually used with scouring powder and water to clean metal surfaces prior to enamelling.

STENCIL: sometimes called a 'template'; made of paper, plastic or silk; used for applying designs to enamel surfaces.

STENCIL BRUSH: a round-nosed bristle brush for pouncing enamel paste through a paper or plastic stencil.

STENCIL KNIFE: a sharp blade with handle used to cut paper stencil.

STILTS: small metal or fired clay pointed bars sometimes used to support an enamel piece while in the furnace.

STRONTIUM SULPHIDE: material used in phosphorescent pigments.

SULPHURIC ACID: used to pickle copper or steel prior to enamelling. The usual combination is 1 part acid : 6 parts water, usually in a warm state. (Always pour the acid into the water and not the water into the acid.)

SYNASOL: trade name for alcohol.

TEMPLATE: a paper, plastic or metal stencil for easy application of designs to enamel surface.

TIMER: a clock that can be set to ring a bell at the time designated for proper firing of enamels when in the furnace.

TONGS: a laboratory appliance for holding hot objects, planches, and screens for transfer in and out of furnaces.

THREAD: thin or thick fired enamel dripping used for textular effects in enamelling.

TRAGACANTH: a hard-flake vegetable gum boiled in water and thinned for use as an enamel adhesive. It should not be used in a thick, heavy state.

TRIVET: metal support made from chrome steel to hold an enamel piece while in the furnace. A non-scaling chromel metal.

VITREOUS ENAMEL: a glass so compounded as to adhere successfully and fuse on to metals, producing a permanent bond. A specific enamel is made for each particular metal for which it is intended.

WET GRINDING: a combination of enamel frit, water, clay and salts ground in a ball mill for 3 to 4 hr to produce 'slush' or 'slip'. Used for dipping or spraying.

21 Materials and Where to Buy Them

BRUSHES

All types from any local art supply company in USA and UK.

COPPER

The American Brass Co, Waterbury, Conn, USA.
Anaconda Brass and Copper (offices in most cities).
A. Boucher (Metals) Ltd, Shelford Place, London, N16, England.
Chase Brass & Copper Co (offices in most cities).
G. Harrison & Sons Ltd, 182 Drury Lane, London, WC2, England.
London Metal Warehouses Ltd, 431 Edgware Rd, London, W2, England.
Henry Righton & Co Ltd, 70 Pentonville Rd, London, N1, England.
Rokker and Stanton Ltd, Gordon St, London, WC1, England.
H. Rollet & Co Ltd, 6 Chesham Place, London, SW1, England.
Rownson, Drew & Clydesdale Ltd, 40 Upper Thames St, London, EC4, England.
J. Smith & Sons (Clerkenwell) Ltd, 42–54 St John Square, London, EC1, England.

COPPER SHAPES AND BLANKS

H. W. Landon & Brothers, 9–12 Bartholomew Row, Birmingham 5, England.

CRUCIBLES (for melting enamel frit)

E. Gray & Sons Ltd, 12–16 Clerkenwell Rd, London, EC1, England.
Morganite Refractories Ltd, Neston, Wirral, Cheshire, England.
Frank Pike Ltd, 58g Hatton Garden, London, EC1, England.

ENAMELS FOR STEEL AND COPPER

C. J. Baines & Co Ltd, Stoke-on-Trent, England.
W. G. Ball Ltd, Longton Mill, Anchor Rd, Longton, Stoke-on-Trent, England.
Blythe Colours Ltd, Cresswell, Stoke-on-Trent, England.
Ceramic Color and Mfg Co, New Brighton, Pa, USA.
B. F. Drakenfeld & Co, Washington, Pa, USA.
E. I. du Pont de Nemours & Co, Wilmington, Delaware, USA.
Enamel-Craft Co, University Center Station, Box 1940, Cleveland 44106, Ohio, USA.
Ferro Corporation, Color Division, 60 Greenway Drive, Pittsburgh, Pa, 15204, USA.
Ferro Enamels Ltd, Ounsdale Rd, Wombourn, Wolverhampton, England.
Harrison Mayer Ltd, Phoenix Chemical Works, Hanley, Stoke-on-Trent, England.
O. Hommel Co, Pittsburgh, 30, Pa, USA.
W. J. Hutton Enamels Ltd, 285 Icknield St, Birmingham, 18, England.
Millenet Co, Geneva, Switzerland.

Pemco Corporation, Baltimore, Maryland, USA.
Schauer & Company, Atzgersdorf St, Vienna, Austria.
American Representative: Norbert L. Cochran, 2540 So Fletcher Ave, Fernindina Beach, Florida, 32034, USA.
Thomas C. Thompson Co, 1539 Old Deerfield Rd. Highland Pk, Ill, USA.
Thompson and Hoseph Ltd, 46 Watling St, Herts, England.
University Circle, 11020 Magnolia Drive, Cleveland, Ohio 44106, USA.
Bernard W. E. Webber Ltd, Alfred St, Stoke-on-Trent, England.
Wengers Ltd, Euturia, Stoke-on-Trent, England.

FOILS (silver and gold)
Johnson Matthey Metals Ltd, 73–83 Hatton Garden, London, EC1, England.
George M. Whiley Ltd, 54–60 Whitfield St, London, W1, England.

FURNACES AND KILNS
Allied Engineering Division, Wombourn, Wolverhampton, Staffs, England.
Allied Heat Co, Ltd, Watford By-pass, Watford, Herts, England.
British Ceramic Service Co Ltd, Wolstanton, Newcastle, Staffs, England.
R. M. Catterson-Smith Ltd, Adams Bridge Works, South Way, Exhibition Grounds, Wembley, Middx, England.
Electric Hotpak Co, Coltman Ave at Melrose St, Philadelphia, Pa, USA.
Ferro Corp, 4150 East 56 St, Cleveland, Ohio, USA.
Ferro Enamels Ltd, Ounsdale Rd, Wolverhampton, England.
Hevi-Duty Electric Co, 4212 Highland Ave, Milwaukee, Wisconsin, USA.
Hoskins Mfg Co, 4435 Lawton Ave, Detroit 8, Michigan, USA.
Kilns and Furnaces Ltd, Keele Street Works, Tunstall, Stoke-on-Trent, England.
L. & L. Mfg Co, 804 Mulberry St, Upland, Chester, Penna, USA.
Pereny Equipment Co, Dept C, 893 Chambers Rd, Columbus, Ohio, USA.
Thomas C. Thompson Co, 1539 Deerfield Rd, Highland Park, Ill, USA.
Bernard W. E. Webber, Alfred St, Fenton, Stoke-on-Trent, England.
James W. Weldon Co, 2315 Harrison Ave, Kansas City 8, Mo, USA.
Wengers Ltd, Euturia, Stoke-on-Trent, England.

GLOVES
At most hardware stores.
The Des Moines Mfg Co, Des Moines, Iowa, USA.
Ferro Corporation, 4150 East 56 St, Cleveland, Ohio, USA.

LUSTRES AND GOLD AND PLATINUM
Blythe Colours Ltd, Stoke-on-Trent, England.
B. F. Drakenfeld & Co, Washington, Pa, USA.
Englehard Industries Division, 1 West Central Ave, East Newark, NJ, USA.
Johnson-Matthey Ltd, Wembley, Middx, England.

METALWORKING TOOLS
Buck & Ryan Ltd, 101 Tottenham Court Rd, London, W1, England.
E. Gray & Sons Ltd, 12–16 Clerkenwell Rd, London, EC1, England.

Herring, Morgan & Southon Ltd, 9 Berwick St, London, W1, England.
Frank Pike Ltd, 58g Hatton Garden, London, EC1, England.
Shetack Tool Works Ltd, 44a Tunstall Rd, London, SW9, England.
S. Tyzack & Son Ltd, 341–345 Old Street, London, EC1, England.

PERFON (metal cleaning solution)
L.R.S. Equipment Ltd, Thame, Oxfordshire, England.

PESTLES AND MORTARS
Bullers Ltd, Milton, Stoke-on-Trent, England.
W. Canning & Co Ltd, Great Hampton St, Birmingham, 18, England.

PLATINUM AND PRECIOUS METALS
Johnson Matthey & Co Ltd, 73–83 Hatton Garden, London, EC1, England.

SPRAY EQUIPMENT, COMPRESSORS AND SPRAY BOOTHS
Binks Manufacturing Co, 3114 W Carroll Ave, Chicago, Ill 60612, USA.
Burgess Products Co Ltd, Electric Tool Division, Sapcote, Leicester, England.
Color Spray Ltd, Albion Works, 1 North Rd, London, N7, England.
Craftools Inc, 396 Broadway Ave, New York 13, NY, USA.
Ferro Enamels Ltd, Ounsdale Rd, Wombourn, Wolverhampton, England.
E. W. Good Ltd, Longton, Staffs, England.
Ohio Ceramic Supply, Ravenna Road, Kent, Ohio, USA.
Passche Airbrushes, Thomas C. Thompson Co, 1539 Old Deerfield Rd, Highland Pk, Ill, USA.
Service Engineers Ltd, Colbridge, Stoke-on-Trent, England.
Sprayon Products Co, 26300 Fargo Ave, Bedford Hts, Ohio, USA.
Thayer & Chandler Airbrush Co (any local art supply co).

STEEL AND SHEET IRON
American Rolling Mill, Middletown, Ohio, USA (offices in most cities).
Bethlehem Steel Co, Bethlehem, Penna, USA (offices in most cities).
Inland Steel Co, 38 South Dearborn St, Chicago, Ill, USA (subsidiaries in most cities).
Richard Thomas & Baldwin Co, RTB House, Gower St, London, WC1, England.
The Steel Company of Wales, Margam House, St James's Square, London, SW1, England.
John Summers & Sons Ltd, St Ermins, Caxton St, London, SW1, England.
United Steel Companies, Rotherham, Yorks, England.
US Steel Co, US Steel Building, Pittsburgh, Penna, USA.

STENCIL PAPER
Multicraft Supply Co Inc, 8617 Carnegie Ave, Cleveland, Ohio, USA.

STILTS AND TRIVOTS (supports to hold enamels for firing)
Artex Mfg, 4038 Huron Ave, Culver City, California, USA.
Atlas Steel Point Stilt Co, 4207 Longshore St, Philadelphia 35, Pa, USA.

Charles H. Draving Co, PO Box 26, Feasterville, Pa, USA.
The Potter's Supply Co, East Liverpool, Ohio, USA.
Thomas C. Thompson Co, 1539 Old Deerfield Rd, Highland Pk, Ill, USA.

TEXTURE MATERIALS

Thomas C. Thompson Co, 1539 Deerfield Rd, Highland Pk, Ill, USA.
University Circle, 11020 Magnolia Drive, Cleveland, Ohio 44106, USA.

TILES (enamel coatings, matt and gloss)

University Circle, 11020 Magnolia Drive, Cleveland, Ohio 44106, USA.

TOOLS AND EQUIPMENT

Ohio Jeweler's Supply Co, 1000 Schofield Bldg, Cleveland 51, Ohio, USA.
Dental tools (for sgraffito drawing).
Any dental supply co.
E. W. Good & Co Ltd, Longton, Stoke-on-Trent, England.

OTHER ENAMELLING SUPPLIERS

Ferro Enamels (Pty) Ltd, PO Box 108, Brakpan, Transvaal, South Africa.
Ferro Enamels Ltd, 16 Bermill Street, Rockdale, NSW, Australia.

22 Magazine Articles by the Author

American Architect Magazine, Jan. 1934.
American Artist, Oct. 1941, Dec. 1941, Sept. 1947, May 1953, Apr. 1956.
American Ceramic Society Bulletin, May 1948, Apr. 1949, Oct. 1952.
American Magazine of Art, Apr. 1933, June 1933, June 1934, Sept. 1934, June 1936, Sept. 1938, July 1939, June 1940.
Architectural Forum, Feb. 1934, Apr. 1941, Sept. 1951.
Architectural League of New York, Gold Medal Award Catalogue 1953, Honorable Mention.
Art Digest and the Arts, Feb. 1954, June 1956.
Art in Focus, Apr. 1952, Feb. 1954.
Art News, Oct. 1936, May 1938, Sept. 1951.
Arts and Decoration, Jan. 1939.
Better Homes and Gardens, Dec. 1947.
Bachelor Magazine, Oct. 1937.
Better Enamelling, May 1935.
Building Research Institute, Nov. 1953.
Business Week, Sept. 1953.
Bystander Magazine, Oct. 1932, Aug. 1933, Nov. 1933.
California Arts and Architecture, Oct. 1936.
Central Glass and Ceramic Research Institute, Calcutta, India, Vol. 4, No. 1, 1957.
Ceramic Age, Feb. 1934, Mar. 1935, Jan. 1941, July 1951, Aug. 1951, Sept. 1951, Feb. 1952, Dec. 1952, Aug. 1953, Sept. 1953, Apr. 1954, May 1954, Jan. 1955, May 1955, Feb. 1956.
Ceramic Industry, Mar. 1934, Sept. 1940, June 1941, Feb. 1945, Aug. 1945, Aug. 1946, June 1947, Sept. 1947, July 1950, May 1951, Oct. 1951, Jan. 1952, Apr. 1952, July 1952, Dec. 1952, Oct. 1953, May 1954, Apr. 1955, Apr. 1956.
Church Management, Apr. 1956.
Clevelander Magazine, May 1935, June 1937, Mar. 1950.
Commercial Art and Industry, London, England, Mar. 1935.
Country Life and the Sportsman, Dec. 1937.
Crockery and Glass Journal, Feb. 1940.
Cross Country Craftsman, Oct. 1951, Dec. 1955.
Decorative Art Golden Gate Exposition, 1939.
Design Magazine, June 1933, June 1935, Nov. 1936, Feb. 1937, Apr. 1937, May 1937, Nov. 1937, Feb. 1942, Apr. 1947, Apr. 1948, May 1950, Oct. 1951, Dec. 1951, Jan. 1952, Dec. 1953, Jan. 1954, Feb. 1954, Nov. 1954, Mar. 1955, Apr. 1955. Nov. 1955, May 1957.
Du Pont Magazine, June 1954.
Emaillerie Magazine, Paris, France, Oct. 1935, May 1936.
The Enamelist, May 1933, June 1933, Aug. 1933, Sept. 1933, Mar. 1934, Apr. 1934, July 1934, Nov. 1934, June 1935, Jan. 1936, May 1936, July 1936, Mar. 1937, June 1938, Sept. 1938, Jan. 1939, May 1939, Feb. 1940, May 1941, July 1941, Oct. 1941, June 1947, Jan. 1949.
Engineering Experiment Station News, Oct. 1947.
Finish Magazine, June 1945, June 1946, July 1947, Jan. 1948, June 1948, Apr. 1948, Aug. 1948, May 1951, Oct. 1951, May 1952, Oct. 1953, Aug. 1955.
Gift and Art Buyer, Mar. 1940, May 1940, June 1940, Oct. 1940, Nov. 1940, Aug. 1941, Feb. 1947, May 1947, June 1947, Aug. 1947, Feb. 1948, June 1948, July 1948, Jan. 1951, July 1951, Dec. 1952, Oct. 1953, Apr. 1954, May 1954, July 1955, Jan. 1957, July 1957, Oct. 1957, Jan. 1958.
House and Garden, Nov. 1940.
Industrial Finishing, July 1934.
Interiors, Aug. 1953.
Iron Age, Feb. 1934, July 1941.
Jewelers Circular Keystone, May 1940.
London Studio, 1932.
Los Angeles Times Magazine, Jan. 1936.
Materials and Methods, Dec. 1953.
McGraw Hill Digest, Sept. 1952, Dec. 1953.
Modern Homes and Gardens, Jan. 1935.
Modern Lamp Accessories, June 1952.
Modern Metals, Nov. 1936, Dec. 1937, Dec. 1950, Oct. 1952.
New Yorker, Dec. 2, 1950, Dec. 7, 1957.
New York Times Magazine, July 1934, Oct. 1937.
Office Management and Equipment, Nov. 1949, Jan. 1952.
Pacific Coast Ceramic News, Feb. 1955, Sept. 1955, Oct. 1955.
Pictures on Exhibit, June 1955, June 1956.
Plain Dealer Pictorial Magazine, May 1937, Nov. 1951, Sept. 1953, Jan. 1955.

Popular Mechanics, May 1933, July 1941.
Popular Science, Dec. 1933, May 1954.
Porcelain Enamel Art for Beginners, Apr. 1947.
Pottery and Glass Salesman, Oct. 1941.
Public Works Magazine, Jan. 1958.
Screen Process Magazine, Nov. 1953, May 1955.
Signs of the Times, Sept. 1936, Aug. 1941. Sept. 1941, Dec. 1941, May 1944, Mar. 1948, Nov. 1951, Jan. 1953, Dec. 1953, May 1955, Apr. 1956.
Steel Magazine, Apr. 1932, Apr. 1934, June 1941.
Steelways, Mar. 1948.
Syracuse Herald Magazine, Nov. 20, 1933.
Today Philadelphia Inquirer Magazine, Dec. 9, 1951.
Today's America Magazine, Mar. 1934.
Town Tidings, Oct. 1932, Nov. 1932.

Magazine Articles By and About the Author

School Arts, 'Crown Filtration Plant Mural', April, 1955.
Architectural Metals, 'St. Mary's Romanian Church Enamels', December, 1960.
Ferroscope, 'St. Mary's Romanian Church Enamels', December, 1960.
Ferroscope, Syd Vickery and enamel mural, October, 1960.
Lead Industries, 'Enameled Aluminum', May, 1960.
Church Management, 'Romanian Church Enamels', Cover & Story, April, 1961.
American Ceramic Society Bulletin, 'Royal Society Appointment of Winter', March, 1961.
Ceramics Monthly, 'Romanian Church Murals', January, 1961.
American Artist, 'St. Mary's Romanian Church Enamels', January, 1961.
Philadelphia Art Alliance Bulletin, 'Winter's Enamel Exhibition', March, 1961.
The International Enamelist, 'Church Murals', August, 1961.
Book Buyer's Guide, 'Article on Beginner's Book', March, 1962.
Lead Industries, 'St. Mary's Romanian Church', Vol. 26, 1962.
International Journal of Religious Education, 'Winters and the Religious Arts Festival at Church of the Covenant'.
School Arts, 'Winebrenner Reviews Winter's Beginner's Book', May, 1962.
Ceramic News, 'Winter's Book Review and Romanian Church', April, 1962.
American Artist, 'Beginner's Book', March, 1962.
Chase Brass & Copper 'Centaur Magazine', Enamel on Copper by Winter, July, 1962.
The Studio Magazine, London, England, 'Review of Beginner's book', August, 1962.
East Ohio Gas, 'Art for the Ages', Romanian Church, Fall, 1961.
Lead Industries Publication, Vol. 26, #2, September, 1962.
Ceramics Monthly, 'Suffolk Museum Exhibit', November, 1962.
Pictures on Exhibit, 'Winter's Enamel Art', November, 1962.
The Arts Features Winter Exhibition, November, 1962.
The Studio, London, England, 'Beginner's Book', November, 1962.
Lead Industries Association, Vol. #4, 'Enameled Restaurant Front', 1962.
The Arts, 'Suffolk Museum Winter Exhibition', December, 1962.
New York Times, 'Suffolk Museum Exhibition', Sunday, November 11, 1962.
Lead Industries, Vol. 26, #4, 'Suffolk Museum Show', 1962.
Popular Ceramics, 'Enameling on Steel', February, 1963.
Metal Products Mfg., 'Full Page Winter Art from Lead Industries', April, 1963.
Christian Art, 'Winter's Lady of Lourdes Enamel', May, 1963.
Ceramic Industry, 'Mott Building, Internal Revenue Mural', April, 1963.
Christian Art, 'Our Lady of Lourdres Enamel', June, 1963.
American Pen Woman, 'Thelma and Ed Winter', October, 1963.
Architectural Metals, 'St. Mary's Romanian Church Murals', October, 1963.

Decorative Art Book, London, England, 1963-4.
Ceramic Industry, 'Precious Metals in Architecture', June, 1964.
Ceramic Industry, Abstract Impressionist Enamel Panel in Color on Cover, August, 1964.
Popular Ceramics, 'Enameling Without a Kiln', November, 1964.
Anno Domini, 'Making an Enamel', May, July and November issues, 1964.
Popular Ceramics, Beginning of 'Questions & Answer Series', December, 1964.
Building Picture, ARMCO, St. Mary's Romanian Church Murals, color, 1965.
Popular Ceramics, 'Painting with Enamels', October, 1965.
Canadian Clay & Ceramics, 'Porcelain Enameling as an Art', Feb.-March, 1965.
Decorative Art Book, Copper Enamel Bowls, 1964-5.
Cleveland Plain Dealer, Bishop's Coat of Arms, June 20, 1965.
The Moravian, 'Art & Religious Experience', January, 1964.
Ceramic Industry, 'Gold Silver Sparkles Porcelain Enamel Panels', June, 1964.
Glass Technology, Sheffield, England, Enamel Book Review, August, 1964.
International Enamelist, Vol. 14 #3, 'Gold and Silver Sparkle Enamels', 1964.
Building Picture, (A.I.A. file) U.S. Government Great Seal Mural, 1965.
Decorative Art, Decorative Accessories in Enamel, 1966-7.
San Diego Evening Tribune, San Diego State College, Winter Enamel, Nov. 4, 1966.
DESIGN, 'Firing Glass Enamel on Aluminum', March-April, 1967.
Syndicated Column, 'Life Begins at Forty', June, 1966.
American Ceramic Society Bulletin, 'Enameling Jewelry', February, 1967.
Studio International, Enamel Steel Panel, May, 1967.
Water and Sewage, 'Porcelain Enamel Murals by Winter', October, 1967.
Studio International, St. Mary's Romanian Church Mural, October, 1967.
Design, 'How to Make a Copper Enamel Bowl', October, 1967.
St. Joseph, Focus on the Arts, 'Most Creative Couple', October, 1967.
Popular Ceramics, 'Enameling Copper', December, 1967.
Popular Ceramics, 'The Enameler's Tools', September, 1967.
Popular Ceramics, 'Painting Enamel Tiles', January, 1968.

23 Bibliography

Addison, Julia De Wolf, *Arts and Crafts in the Middle Ages*, L. C. Page & Co, Boston, 1908.

Andrews, A. I., *Enamels*, Garrard Press, Champaign, Illinois, 1936.

Bates, Kenneth F., *Enamelling, Principles and Practice*, World Publishing Co, Cleveland, 1951.

Biegeleisen, J. I. and Max Arthur Cohn, *Silk Screen Techniques*, Dover Publications, New York.

Brown, W. N., *The Art of Enamelling*, Scott, Greenwood & Co, London, 1900.

Bryant, Eugene E., *Porcelain Enamelling Operations*, Enamelist Publishing Co, Cleveland, Ohio, October, 1953.

Cleveland Museum of Art, *Bulletin*, May, 1933.

Cunnynghame, H. H., *The Art of Enamelling Metal*, Archibald Constable & Co Ltd, Westminster, 1899.

De Koningh, H., *The Preparation of Precious and Other Metals for Enamelling*, The Norman W. Henley Publishing Co, New York, 1930.

Encyclopaedia Britannica, *Enamel*, Vol. 8.

Feirer, John L., *Modern Metalcraft*, The Manual Arts Press, Peoria, Illinois, 1946.

Fisher, Alexander, *The Art of Enamelling on Metal*, The Studio, London, 1906.

Hansen, J. E., *A Manual of Porcelain Enamelling*, 1943.

Hart, G. F. and Keeley, Golden, *Metal Work for Craftsmen*, Sir Isaac Pitman & Sons Ltd, London, 1945.

Kosloff, Albert, *Screen Process Printing*, Signs of the Times, Cincinnati, Ohio.

Landrum R. D., *Enamels 1918*, Harshaw Chemical Co, Cleveland, Ohio.

Larom, Mary, *Enamelling for Fun and Profit*, David McKay Co Inc, New York, 1954.

Lehnert, Georg Hermann, ed. *Illustrierte Geschichte des Kunstgewerbeschule*, **1**, M. Oldenbourg, Berlin, 1907.

Martin, Charles J., *How to Make Modern Jewelry*, Simon and Schuster, New York, 1949.

Metropolitan Museum of Art, *Bulletin*, Museums of Modern Art, New York, *Masterpieces of Enamelling*, May, 1940.

Millenet, Louis Elie, DeKoningh H., *Enamelling on Metal*, Translation from the French, Crosby Lockwood & Sons, London, 1926.

Miller, John G., *Metal Art Crafts*, D. Van Nostrand Co Inc, 1949.

Morgan, J. Pierpont, Catalogue of the Collections of Jewels and Precious Works of Art, Chiswick Press, London, 1910.

Neuburger, Albert, *The Technical Arts and Sciences of the Ancients*, Macmillan Co, New York, 1930.

Otten, Mizi, and Berle, Kathe, *The Art of Enamelling Can Be Fun*, 1950.

Pack, Greta, *Jewelry and Enamelling* D. Van Nostrand Co Inc, New York, 1941.

Rossenthal, Rudolf and Ratzka, Helen L., *The Story of Modern Applied Art*, Harper and Brothers, New York, 1948.

Stuckert, L., *Die Emailfabrikation*, 1929.

Thompson, Thomase E., *Enamelling on Copper and Other Metals*, Thomas C. Thompson Co, 1950.

University of Pittsburgh, Department of Fine Arts, *History of Enamels*, Catalogue of Exhibition, April 8, 1950.

Untracht, Oppi, *Enamelling on Metal*, Greenberg Publishers, New York, 1957.

Wiener, Louis, *Hand Made Jewelry*, D. Van Nostrand Co Inc, New York, 1948.

Winebrenner, D. Kenneth, *Jewelry Making as an Art Expression*, International Textbook Co, 1953, Scranton, Pa.

Winter, Edward, *Porcelain Enamel Art for Beginners*, Enamelist, Publishing Company, Cleveland, Ohio, 1947, (Pamphlet).

Winter, Edward, *Enamel Arts on Metals*, Watson-Guptill Publications, New York, 1958.

Winter, Edward, *Enamelling for Beginners*, Watson-Guptill Publications, New York, 1962.

Vargin, V. V., *Technology of Enamels*, Translated from the Russian by Kenneth Shaw, Maclaren and Sons, London, 1967.

Index